A Star Hopper's Guide to Charles Messier's Famous List

110 THINGS
TO SEE WITH A
TELESCOPE

"THE WORLD'S MOST FAMOUS STARGAZING LIST"

John A. Read
with Chris Vaughan

Foreword by Tim Russ

 T0203250

 sourcebooks

For more stargazing tips and tricks, visit:
www.LearnToStargaze.com
@LearnToStargaze
www.youtube.com/c/LearnToStargaze

Copyright © 2021, 2024 by John A. Read and Chris Vaughan
Cover Design by John Read, Jennifer Read
Title text layout from 50 Things to See with a Telescope, Tyler Cleroux, Formac Publishing
Interior design by Jennifer Read/Stellar Publishing
Cover image © Anon/Adobe Stock
Cover foreground image by our good friend Sean MacAully (1978–2018)

This publication is designed to provide accurate and authoritative information in regard to the subject matter covered. It is sold with the understanding that the publisher is not engaged in rendering legal, accounting, or other professional service. If legal advice or other expert assistance is required, the services of a competent professional person should be sought. —*From a Declaration of Principles Jointly Adopted by a Committee of the American Bar Association and a Committee of Publishers and Associations*

All brand names and product names used in this book are trademarks, registered trademarks, or trade names of their respective holders. Sourcebooks is not associated with any product or vendor in this book.

Published by Sourcebooks
P.O. Box 4410, Naperville, Illinois 60567-4410
(630) 961-3900
sourcebooks.com

Originally published in 2021 in Canada by Stellar Publishing.

Cataloging-in-Publication Data is on file with the Library of Congress.

Printed and bound in China.
OGP 10 9 8 7 6 5 4 3 2 1

Foreword

My interest and passion for space science began years ago. I have always been fascinated by the subject and follow the most current discoveries made by astronomers.

I began my exploration of the cosmos with a small Newtonian, wide-field telescope called an Astroscan. At that time there were no affordable, small, consumer computerized telescopes on the market so I had to learn the night sky by using star charts, and star hopping to find objects. Looking back, I'm glad I was able to learn the night sky that way.

I began by looking for the more prominent Messier objects, as those tended to be the brightest and easiest to find. I preferred using the Telrad Reflex Sight Finder, as I have found it's the most practical, reliable, and accurate finder scope. If the optical tube of your scope is too short to mount one, then the Rigel QuikFinder is a good substitute.

I subsequently upgraded to larger Dobsonian and Schmidt-Cassegrain telescopes. I have found that the preferred type and size of telescope depends on what your interests and objectives are. For some, it's astro-photography. For others like myself, it's visual observation. Consumer telescope technology has advanced quite a bit since I started exploring the night sky, and is accelerating every year as new "go-to" computerized telescopes and star chart apps for your smart phones hit the market.

But overall, I've found that the way to go when starting out in astronomy is to keep it simple — with a small to medium-sized, easy-to-use, non-computerized telescope. Dobsonian reflectors are an ideal choice, and the prices are best. And, their design lets you look straight up over head! You won't need to worry about getting to the dark sky site and finding your batteries aren't charged on your computerized scope, or that it's experiencing technical difficulties, and your whole night is a waste. On the other hand, the computerized, "go-to" telescopes do make it easier to find many objects that may be visually difficult to find. The solution is to start small and simple, and work your way up to larger and more complex models as you gain experience.

Over the past few years I have had the pleasure of being able to share views through my telescope with the general public at Griffith Observatory here in Los Angeles. There are times during the viewing sessions when people sincerely express an interest in the hobby, and ask me how to get started. The next time they do, I will simply refer them to this book. In my opinion, it is the definitive guide to amateur astronomy!

Tim Russ
Actor, Musician, Astronomer

TABLE OF CONTENTS

SUMMER TARGETS

Introduction

Chris and I are, first and foremost, visual astronomers. I started taking amateur astronomy seriously about 15 years ago and purchased my first "real" telescope, a Meade ETX Go-To telescope. Chris was given an astronomy book at 10 years old and has been looking up ever since.

From my backyard just outside of San Francisco, I only had a small patch of open sky to the west. I'd point the telescope between the trees and observe whatever objects were in season. I soon realized that I'd see a whole lot more if I had more aperture. So, after a bit of research I ordered a 12-inch Dobsonian telescope.

I'd never used, or for that matter, ever seen a Dobsonian. When the box arrived, I was shocked at its size. Then I realized that box was only the base! An even larger box arrived with the optical tube. It was a crisp January evening when I set up the telescope outside and pointed it at the Orion Nebula, and couldn't believe my eyes!

It was like seeing Saturn for the first time, all over again. I stepped back in amazement and then ran into the house trying to convince everyone to come outside and take a look (with marginal success).

Like most beginner astronomers, I focused primarily on the Messier (pronounced Messy-ay) objects. After about a year, I was starting to repeat the Messier objects I'd first observed. I began volunteering with the Mount Diablo Astronomical Society and the Mount Diablo State Park System. I'd share the night sky with students, driving from event to event with the giant telescope strapped to the roof of my car.

John's 12-inch Dobsonian telescope

Orion Nebula imaged with John's 102 mm refractor telescope

Armed with my telescope and a Telrad finder, I got pretty good at teaching people to star-hop. My favorite objects to share were the Great Globular Cluster in Hercules (M13), the Dumbbell Nebula (M27), the Wild Duck Cluster (M11), and the Butterfly Cluster (M6). I'd even make my own star maps, with the star-hops all planned out. These techniques led me to put together my first book, *50 Things to See with a Small Telescope*, which was an unexpected success.

110 Things to See with a Telescope is the book I wish I'd had back when I upgraded to the Dobsonian telescope. This book is not necessarily geared toward the first-time stargazer, although a new observer would have little trouble using it. Let's say it's written for the second-time stargazer—for someone who, like me, loves to hop from target to target, viewing as many objects as possible in a single evening.

Finding the 110 objects from Charles Messier's list is a milestone for the amateur astronomer. For this reason, organizations such as the RASC (Royal Astronomical Society of Canada) and the Astronomical League (in the USA) offer certificates and pins to observers who have found and recorded their visual observations of the objects in this list. We'll guide you though the list and let you track your progress. With the help of this book, you'll have your Messier certificate before you know it.

If you're planning a Messier Marathon (attempting to see all 110 objects in one night), this book will guide you through that process, too. Whether you're using a small telescope under dark skies, or a larger scope in the suburbs, we've done our best to set you up for success!

~ John A. Read

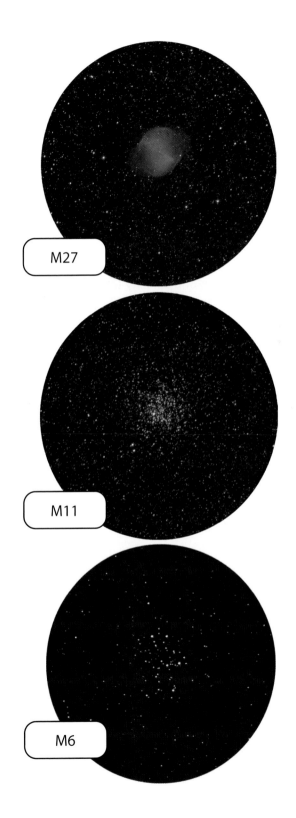

M27

M11

M6

Introduction

History of the Messier List

"Objects that are not comets"

In the mid-to-late 1700s, a French comet hunter named Charles Messier created a list of 103 fuzzy objects he saw through his telescope. At the time Messier didn't know what those "imposters" were, but he knew that they weren't comets because they didn't change position from night to night. We now know that the objects in his list are star clusters, nebulae, and galaxies. Today, those deep-sky objects are labeled on star charts and astronomy apps using Messier's initial "M" plus the number he assigned them in his catalog.

Charles Messier at around age 40

Hôtel de Cluny, Paris

Long after his death in 1817, notable astronomers pored over the notes left by Messier and a fellow astronomer and friend named Pierre Méchain. Based on a detailed review of those documents, seven additional objects have been added to Messier's list: Camille Flammarion added M104 in 1921; Helen Sawyer Hogg added M105, M106, and M107 in 1947; Owen Gingerich added M108 and M109 in 1953; and Kenneth Jones added M110 in 1966.

It's important to note that Charles Messier did almost all of his work from his observatory at Hôtel de Cluny in Paris, France (now a museum). For this reason, many of the Messier objects are positioned too far north in the sky to be seen from most of the Southern Hemisphere. In fact, only those who observe at latitudes between 20 degrees south and 55 degrees north can see all the Messier objects.

Charles Messier's Telescope

In his quest for comet-hunting fame, Charles Messier used a variety of telescopes. Some of them had apertures of over 6 inches (152 mm) and focal lengths as long as 30 feet (9 m). However, the mirrors in his telescopes were made of polished metal and were not nearly as reflective as modern glass mirrors. Glass telescope mirrors would not be invented until 1864, almost 200 years after Isaac Newton built the first reflector telescope! Today's 8-inch Dobsonian telescopes (the optimal beginner telescope) are far superior to any of the instruments Charles Messier would have used.

Having a better telescope than Charles Messier doesn't mean your task will be easy. Light pollution from cities and towns is a challenge that eighteenth century astronomers never had to face. It's also possible that our collective addiction to phones and computer screens degrades our eyes' ability to see dim objects.

Gregorian Reflector

It is said that of all the telescopes Charles Messier used, his favorite was a Gregorian Reflector. These telescopes had long focal lengths but were compressed into a small tube by the use of two concave metallic mirrors.

What Are the Messier Objects?

Not Comets!

For the most part, all that Charles Messier knew about his objects was that they were not comets. Astronomers of the day referred to most of them as nebulae, which simply meant "cloudy." We now know that these objects are made primarily of stars, gas, and dust—in the form of galaxies, nebulae, and star clusters. Charles Messier's list also contains a double star (M40), a dense patch of stars (M24), and an asterism (M73).

Galaxies

When we look at a galaxy through our telescopes, we're seeing the combined light from hundreds of billions of stars, plus glowing gas and reflective dust, all located millions of light-years away. Galaxies generally fall into a few basic categories: spiral, elliptical, lenticular (which are spiral galaxies without visible structure), and irregular. The Messier list contains 40 galaxies.

Andromeda [Spiral] Galaxy (M31) imaged with a 102 mm refractor telescope

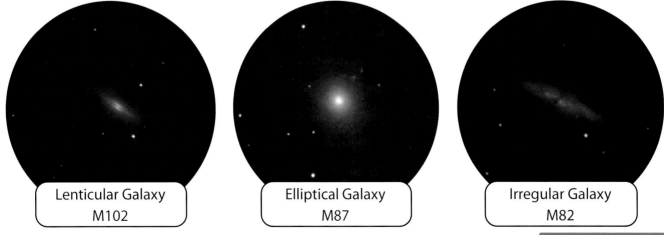

Lenticular Galaxy
M102

Elliptical Galaxy
M87

Irregular Galaxy
M82

Open Star Clusters

An open cluster is a group of young stars all born from the same collapsed cloud of gas and dust. Over time, the individual stars in the cluster disperse, perturbed by our galaxy's gravity. Open clusters are part of our Milky Way Galaxy. The closest ones can appear huge! The Messier list contains 26 open clusters.

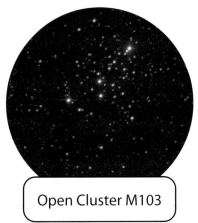

Open Cluster M103

Globular Star Clusters

Globular clusters are tight groups of hundreds of thousands of stars, bound into a round shape by gravity and orbiting our galaxy in a region called the Halo. These clusters contain very old stars, and some may host black holes. The Messier list contains 29 globular clusters.

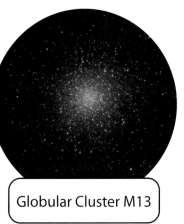

Globular Cluster M13

Nebulae

There are several types of nebulae in the Messier list. Planetary nebulae occur when a dying star puffs off its outer layers. Supernovae remnants are the remains of an exploded star. Emission nebulae are clouds of glowing gas, while reflection nebulae are primarily illuminated dust. Though colorful in photographs, these objects typically appear gray to the human eye. The Messier list contains 12 nebulae.

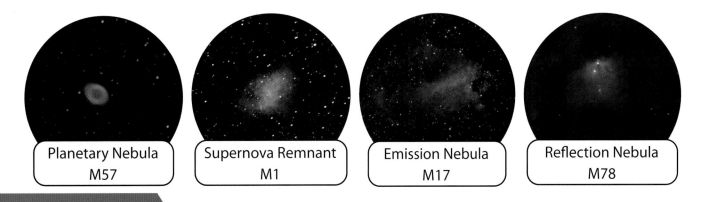

Planetary Nebula
M57

Supernova Remnant
M1

Emission Nebula
M17

Reflection Nebula
M78

Effective Stargazing

Plan Your Evening

Before you go out under the stars, review the list of targets you plan to view. Objects in the west set first, so you may want to observe these first. Dimmer objects may require a darker location, and moonless skies.

You'll notice that the objects in this book are not listed in Messier's numerical order. That's because they're listed in the most effective viewing order for each season. This order will also help you efficiently complete observing programs and marathons.

Red light headlamp

Adapt Your Eyes

The Messier list is composed entirely of deep-sky objects, or DSOs. Most of the objects are quite dim. In order to see them more easily, allow your eyes to adapt to the dark by refraining from looking at a phone or computer screen, car headlights, or porch lights for at least twenty minutes. Dim red light will not harm your night vision, so use a red light for reading this book outside and while taking notes at the telescope.

Become "Moon Aware"

Even if you're observing far from city lights, moonlight will brighten the sky and obscure many of the fainter targets. Be aware of the lunar phase and location of the Moon in the sky. Even during the crescent phase, objects near the Moon may be obscured.

Ideally, searching for Messier objects is done on evenings when the Moon is not in the nighttime sky at all, generally from third quarter until a few days after New Moon.

The RASC Observer's Calendar includes the Moon phase for every night of the year.

Location Matters

We do most of our stargazing from our backyards within city limits, so we have to be realistic about what we're able to see. Even with large telescopes (like a 12-inch Dobsonian), we cannot see all of the objects in this book from a city or large town.

However, from a park thirty minutes away from the city, even much smaller telescopes, like 4-inch refractors, are capable of viewing the dimmer Messier galaxies.

For those of us living in the Northern Hemisphere, it also helps to set up our telescopes where there is a clear view to the south. As the Earth rotates, this is where the seasonal targets climb highest in the sky.

Isaac Read using his 5-inch Newtonian

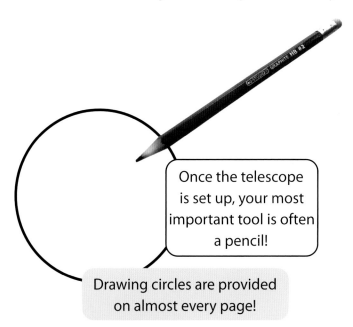

Once the telescope is set up, your most important tool is often a pencil!

Drawing circles are provided on almost every page!

Write It Down!

One of the most fulfilling parts of this hobby is tracking your progress. By looking back over your notes and sketches, you'll realize just how much there was to see in the night sky. Don't forget to include your co-observers in your notes to keep those memories alive!

Several organizations (including the RASC and the Astronomical League) will award you certificates for your observations, but only if you write them down! That's why each page in this book includes a place to record the details of your observation, and a circle for sketching what you see. Application instructions are also provided at the end of this book.

Seeing

Even though the sky may be free of clouds, other factors, such as humidity, dust, wind, and temperature, can affect your ability to see Messier objects.

"Seeing" measures how sharp and steady objects appear in the eyepiece. It is measured by analyzing the diffraction pattern of a bright star. Poor seeing is caused by high-altitude air turbulence. In general, the larger and more twinkly the stars appear, the worse the seeing conditions. Note that seeing conditions may change during your observing session.

Images of a star's light under various "seeing" conditions. Seeing is rated between 1 (best) and 5 (worst).

Transparency

Transparency is a measure of how transparent the atmosphere is. Imagine trying to use a telescope with the lens covered in dew. When there is a lot of humidity in the air, the transparency is reduced.

Transparency is also affected by your elevation (observing at higher elevation is better), smoke or dust, and atmospheric pressure.

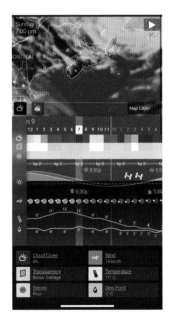

Having good seeing and transparency is essential for optimal views of deep-sky objects—particularly galaxies and dim nebulae.

Astronomers often estimate transparency by noting the magnitude of the dimmest star they can see without a telescope.

We See "Airy Disks," Not Stars

The stars in our night sky are so distant that they have no apparent diameter. So why do brighter stars appear larger? It has to do with how the light behaves when it enters the telescope. Stars appears as tiny disks, called Airy disks, after an astronomer named Sir George Biddell Airy. The larger the aperture of the telescope, the smaller the size of the Airy disks. The brighter the star, the more of the star's Airy disk we are able to see. Resolution is defined by the minimum visible separation between two side-by-side Airy disks.

An app called **Astrospheric** forecasts the "seeing" and "transparency" at your location, hour by hour.

Finding the "Dark"

Amateur astronomers talk a lot about the darkness of their skies. You may hear them refer to the "Bortle" scale. In this system, 1 represents a perfectly dark location, 5 is the suburbs, and 9 is the inner city.

Astronomers also use light pollution maps to choose where to stargaze. You can see most of the Messier objects from a green zone, but to find and explore the dimmest galaxies, a blue or gray zone is highly recommended.

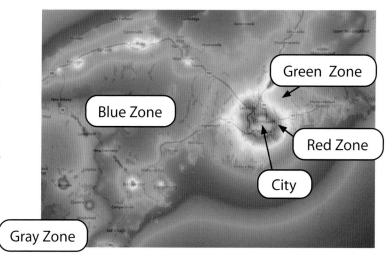

Use a website like https://darksitefinder.com/maps/world to find the darkest skies nearest you.

Right-clicking on the light pollution map will usually reveal the Bortle Class value for that location.

Averted Vision

Many of the Messier targets shine at the limit of human vision, even in larger telescopes under dark skies. Use the averted vision technique to brighten faint objects. Instead of looking directly at the object, cast your gaze toward the edge of the eyepiece field of view. That lets the light-sensitive receptors in your eye brighten the object!

The Tap Tap Trick

A final trick for revealing a very faint target that should be visible but isn't—or for better seeing the fine details in a deep-sky object, like a galaxy's spiral arms—is to gently tap the telescope while you are looking in the eyepiece. This forces the light coming through your telescope to hit slightly different parts of your eye, and also uses the wiring in your brain that keys on motion.

Star-Hopping

Star-hopping is the process of finding targets near recognizable patterns in the sky. The most common example would be using the "pointer" stars to find the North Star (a.k.a. Polaris). Sometimes we'll describe star-hops with words—at other times, by a dashed arrow on the provided sky maps.

Star-hopping is generally a naked-eye technique, but you can follow star patterns through a finderscope or telescope as well. When you are observing the Messier targets in the Virgo Galaxy Cluster, where there are more galaxies than stars, you'll find yourself "galaxy-hopping" using the view in your eyepiece.

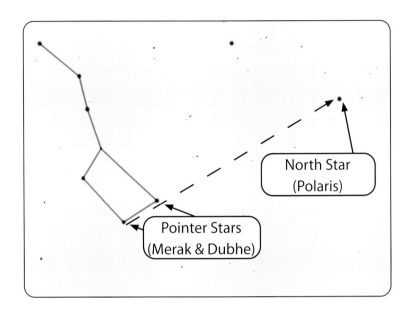

North Star (Polaris)

Pointer Stars (Merak & Dubhe)

Finderscope

Telrad

Finderscope vs. Unit Finder

The narrow view through your telescope makes it hard to find objects, so you're going to need a finder. Finders can apply some magnification to the sky, or none at all. Either type must be precisely aligned with the telescope, so that the finder and the telescope are pointed at exactly the same spot. The alignment can be done in daytime using a distant landmark.

A finderscope is a small telescope with a wide field of view, usually with cross-hairs in the center. These work well for bright objects like planets and bright stars, but from the city, getting a deep-sky object to appear in your finder can be a challenge.

Instead of magnifying the sky, unit power (or 1x) finders project a bullseye (or dot) onto a tiny window. While sighting through the window, move the telescope until the bullseye is over the object's precise location based on the map. Assuming your finder is correctly aligned, this will center the target in the eyepiece.

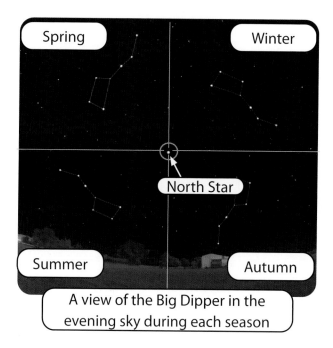

A view of the Big Dipper in the evening sky during each season

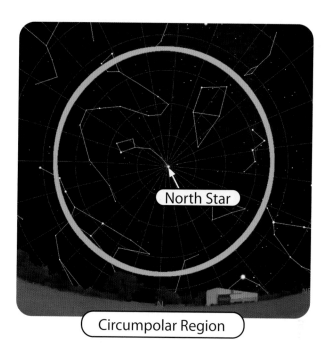

Circumpolar Region

Seasonal Skies

Our view of the stars changes based on Earth's position in its orbit around the Sun, and on the time of night. Although the Messier objects in this book are organized by season, at least half them are visible on any given night. Want to see the winter sky during autumn? Just wait a few hours after sunset and watch the winter constellations rise in the east.

In the autumn, the earlier sunsets delay when constellations in the western sky sink into the twilight. That means that many of the summer objects will still be visible during autumn. In fact, objects near the Summer Triangle are visible until December!

In the spring, the opposite is true. The days are getting longer and the nights are getting shorter. Some springtime objects, like the Beehive Cluster (M44), become lost in the sunset before the end of the season.

Circumpolar Targets

As the Earth rotates on its axis (and orbits the Sun), there is a circular zone of the sky, called the Circumpolar Region, that never sets below the horizon. From the Northern Hemisphere, this region surrounds Polaris, the North Star. In the Southern Hemisphere, stars near the constellation Octans are circumpolar.

The size of the Circumpolar Region depends on your latitude. If you live at 25 degrees north latitude, then only the stars within 25 angular degrees of the North Star will be circumpolar. The stars inside that ring are in your sky all night and all year.

Although circumpolar targets can be viewed on any night of the year, we've assigned them to the season when they're highest above the horizon in the evening.

Know Your Telescope

Successfully observing all the Messier objects requires you to know your telescope. Can you point it accurately enough to see objects invisible to the naked eye? Does it have a large enough aperture to show faint objects and to resolve intricate details? What is the appropriate magnification? Should you use a Barlow? Or a filter?

Aperture and focal length are generally written on the telescope. Eyepiece focal length is written on the side of each eyepiece. Those numbers can be used to calculate magnification, resolution, focal ratio, and more.

This Dobsonian's aperture is 305 mm (12 inches) and its focal length is 1,500 mm.

Magnification

In general, it's easier to find deep-sky objects using low magnification because you'll be viewing a larger patch of sky. Only once you've located an object will you increase the magnification. Be sure to experiment with different eyepieces. For example, if you "zoom in" on a nebula with a high-powered eyepiece, it will generally appear fainter because fewer photons (bits of light) will hit each photorecepter in your eyes.

Subjectively, more magnification can improve the view of some objects. This is because as you zoom in, the background becomes blacker. Although the object dims too, your eye may have an easier time picking out details against the darker background.

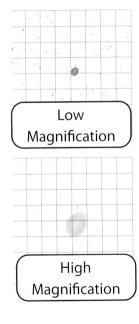

Low Magnification

High Magnification

Barlows fit between the eyepieces and the telescope. Barlows typically double or triple the magnification.

$$\text{Magnification} = \frac{\text{Telescope Focal Length}} \div \frac{\text{Eyepiece Focal Length}} \times \frac{\text{Barlow Strength}}$$

Remember: When doing math, always make sure your units match!

Resolution

Resolution, or resolving power, is a telescope's ability to see fine details on objects in space. A telescope's resolution is determined entirely by aperture, the diameter of the primary mirror or lens. Larger aperture telescopes offer superior resolution.

Resolution is measured by the smallest angle a telescope can distinguish between two objects (such as two stars). Note that resolution is limited by the seeing conditions of the atmosphere. There are several ways to calculate resolution, but one of the most common is called the **Rayleigh resolution**, which simplifies to the following:

Resolution (arc seconds) = 138 ÷ Aperture (mm)

Focal Ratio

In simple terms, the focal ratio compares a telescope's length to its diameter. A high focal ratio, or "slow" telescope, has focal ratio of around 10 or higher. These telescopes are best for viewing planets at high magnification.

A "fast" telescope, such as an f/4 or f/5 Newtonian reflector or Dobsonian, offers a much wider field of view. Low focal ratio telescopes with large apertures are the best ones for viewing deep-sky objects.

Focal Ratio = Focal Length ÷ Aperture

Maximum Useful Magnification

Attempting to search for—and view—objects at high magnification is a common mistake made by beginners. There's a limit to how much you can "zoom in" before a telescope's limitations (and seeing conditions) spoil the view, even for easy targets like Jupiter or the Moon.

The Maximum Useful Magnification can be estimated by doubling your telescope's aperture in mm (or by multiplying the aperture in inches by 50). For example, a telescope with a 70 mm aperture has a maximum useful magnification of 140x, and an 8-inch telescope has a maximum useful magnification of 400x.

How to Focus Your Telescope

Center a bright star in your eyepiece. At first, it may look like a giant circle or donut. Adjust the focusing knob until the star is small and sharp.

How to Align Your Finder

In order to find things in the sky, your finder must be aligned! This is easier during the day. Point the telescope at a distant object like a chimney. Adjust the knobs on the finder until both the telescope and the finder are pointed at exactly the same spot.

Filters

The light we see from nebulae is composed of specific wavelengths (colors) that are generated when the nebula's gases are energized by the radiation emitted from nearby stars.

Special filters can make nebulae more visible by allowing only specific wavelengths to pass through your telescope, blocking certain wavelengths containing light pollution. Narrowband Ultra High Contrast (UHC) and Oxygen-3 (O-III) filters work particularly well on emission and planetary nebulae, and can also improve views of reflection nebulae. Broadband Light Pollution Reduction (LPR) or Skyglow filters generally increase contrast for nebulae, galaxies, and clusters by darkening the sky around them.

If you're purchasing filters, be sure to match the filter diameter to your eyepiece's barrel (usually 1.25 inches). Screw the filter onto the bottom of the eyepiece, then insert the eyepiece into the telescope. Red, Green, and Blue filters are generally used with monochrome cameras to create color images.

Narrrowband filter

Alt-Azimuth Mounts

Alt-azimuth (Alt/Az) mounts are the best choice for beginners. They let you stand behind the telescope and move it around the sky without restriction. Premium mounts offer slow motion knobs to let you follow the objects as they move across the sky.

If you're a beginner stargazer working through this book, we highly recommend you start with a Dobsonian-style telescope. The Dobsonian mount features a lazy Susan, which lets the telescope sweep left and right, and a rocker that lets the telescope tilt up and down.

Refractor telescope on a Twilight Alt-azimuth Mount

8-inch Dobsonian telescope

Equatorial (Eq) Mounts

Equatorial (Eq) mounts are a poor choice for beginners because they need to be aligned to the celestial pole and be weight-balanced. They are typically used for astrophotography because the mount accounts for the way objects rotate across the sky during the night.

How to Align an Equatorial Mount

You **must** point this axis at the celestial pole (if you live in the Northern Hemisphere, this is near the North Star).

This is the **declination** control. It moves your telescope north and south, away from or toward the North Star.

This is the **right-ascension (RA)** control. It moves the telescope east and west, opposite to the rotation of the Earth.

Release this lock to tilt this axis until it points at the celestial pole. The reading on the dial should match your latitude.

Position the counterweight so that the telescope is balanced throughout its range of motion. (It stays put when you unlock the RA axis.)

Loosen this lock to move the mount left and right during the alignment process. Better mounts often have two left-right fine-adjustment knobs as well.

Celestron C8 on an Advanced VX Go-To Mount

Go-To Telescopes

If you're observing the Messier targets as part of an official observing program, be sure to check the rules to see if Go-To telescopes are allowed.

While able to find and track objects automatically, Go-To telescopes can be expensive, and a challenge for beginners to master. We recommend spending that extra money on a larger aperture manual telescope.

That said, a reasonably large aperture Go-To telescope, precisely aligned to a dark sky, will make quick work of the Messier objects.

Messier Marathons

Messier Marathons are generally carried out in the middle of March, on a night when the Moon is closest to its new phase. On these nights, stargazers from all over challenge themselves to see as many Messier objects as possible, in a single night.

These stargazing marathons happen in March because that's when the Sun is located where there are no Messier objects. Objects low in the western sky are viewed first, and objects rising early in the morning to the east are viewed last.

The targets in this book are ordered by season. However, the optimal order for a Messier Marathon is slightly different. Some autumn targets are viewed just after sunset, and other autumn targets just before sunrise (see page 1 for details).

Often, stargazers will complete the Marathon in two stages. During the few hours after sunset they observe 60 or so targets, then they go to bed. They wake up a few hours before sunrise and observe the remaining 50 or so Messier objects.

History

In the late 1960s to mid-1970s, several amateur astronomers realized that it might be possible to observe all Messier objects in a single night.

Beginning in 1977, the first attempts were made to do just that. The original "Messier Marathon" can be traced to four astronomers, Tom Hoffelder, Tom Reiland, Ed Flynn, and Donald Machholz.

In 1979, amateur astronomer and author Walter Scott Houston wrote about the Messier Marathon in *Sky & Telescope* magazine, and its popularity grew from there.

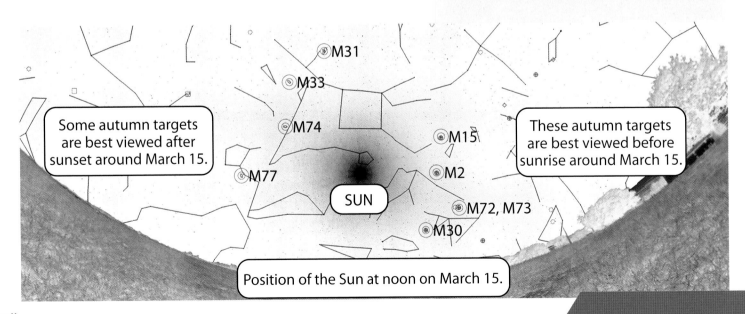

Some autumn targets are best viewed after sunset around March 15.

These autumn targets are best viewed before sunrise around March 15.

Position of the Sun at noon on March 15.

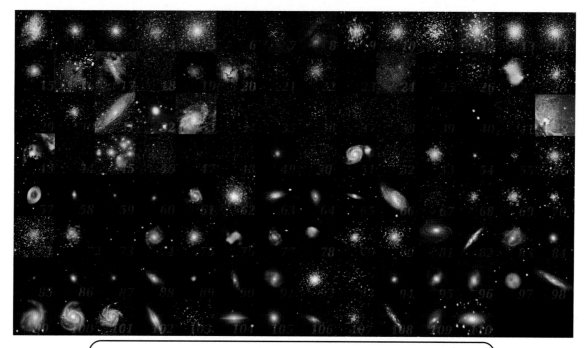

Images of all Messier objects. Credit: Michael A. Phillips

Modified Marathons

Messier Marathons don't have to be restricted to March. We both do most of our stargazing in the evening and would much prefer a series of "mini marathons," scheduled once per season. With this method, from dark skies, all the Messier objects can be observed over the year, without rushing or staying up all night. John recently participated in a robotic Messier Marathon, thanks to some pretty cool technology. His university's telescope has been modified so that it can be remotely controlled via the Internet in real time!

Certificate Programs

Several astronomical societies around the world offer certificates to members (and sometimes non-members) for observing all or most of the Messier objects.

Generally, participants are required to record their observations, including details such as the object, date, time, sky conditions, telescope used, etc. They are often encouraged to draw how the object appears through their eyepiece. That's why an observing log and drawing circle have been added for every object in this book.

Both the Royal Astronomical Society of Canada (RASC) and the Astronomical League offer several popular certificate programs, including the Messier List.

The RASC's Messier Pin

Using This Book

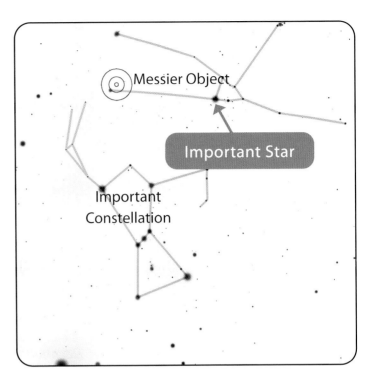

Maps: Each page in this book should be sufficiently detailed for finding the associated Messier object. Each map features a prominent constellation or asterism (star pattern) for the season. For example, winter targets tend to feature Orion, spring targets feature Leo, and summer targets feature either the Summer Triangle asterism or the Teapot asterism. Only those objects relevant to finding your targets have been highlighted or labeled.

Telrad Rings: Each map features what's called a Telrad* ring (a bullseye) over the Messier object. The inner ring represents 1/2 degree of sky (or a magnification of about 100x), while the outer rings represent 2 degrees and 4 degrees (25x and 12.5x), respectively. These values are assuming an eyepiece with a field of view of approximately 50 degrees.

Using a Telrad Ring

The maps in this book are vastly easier to use if you are using a finder that does NOT magnify the sky. These include Telrads, QuikFinders, and Red Dot Finders.

Simply align your finder to the exact location shown by the (⊙) symbol on the map.

If you're using a finderscope, it may help to use a star atlas (like *The Cambridge Star Atlas*) in addition to our maps.

* Telrad is a brand of Unit Finder.

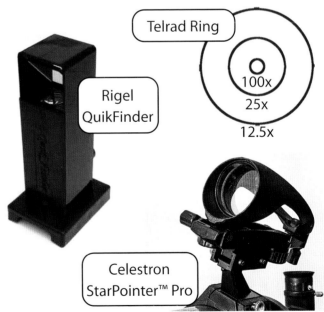

The QuikFinder and StarPointer Pro are also bullseye finders.

Eyepiece View: The simulated eyepiece images show photos of an area of sky slightly smaller than that of the full Moon. These images simulate approximately 100x magnification, unless otherwise noted. To see the galaxies (the dimmest targets in this list) exactly the way they appear in these images, you'll need: moonless nights, perfectly dark skies, and a moderate to large aperture telescope.

Magnitude: For each object we've noted the visual magnitude, which is a measure of brightness. The higher the magnitude value, the dimmer the object. Brighter objects are not necessarily easier to see. If an object is "extended," like a galaxy, its light will be distributed across a larger area of sky, rendering it fainter overall. In those cases, we'll mention the "surface brightness" (brightness per unit area) as a measure of how visible the object will appear given certain observing conditions.

Additional Information: We have provided the formal common names, or informal nicknames, for each object. In the text, we talk a little about Messier's, and our own, experiences with the objects, and describes how best to find them. We've also suggested some things to look for in each object. But not too many—we want you to discover and record your own impressions!

Difficulty: We have suggested the level of difficulty for each target. This is mostly a measure of how hard it was for **us** to find each object, so it is largely subjective. 1 = Easy. 3 = Challenging, but visible under most conditions. 5 = Requires dark skies and perfect seeing conditions.

Measuring Sky Brightness

The sky's brightness can also be measured using surface brightness (in units of magnitude per square arcsecond). If an object's surface brightness is more than a few magnitudes dimmer than the sky, the object is effectively rendered invisible to the human eye. A rough estimate for sky surface brightness is 17 for the city, 19 for the suburbs, and 21 for the countryside.

Images are gray-scale inverted for easier nighttime viewing.

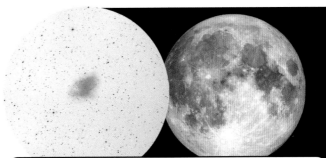

Almost all of the Eyepiece View images are scaled to 100x, or about the size of the full Moon.

AUTUMN TARGETS

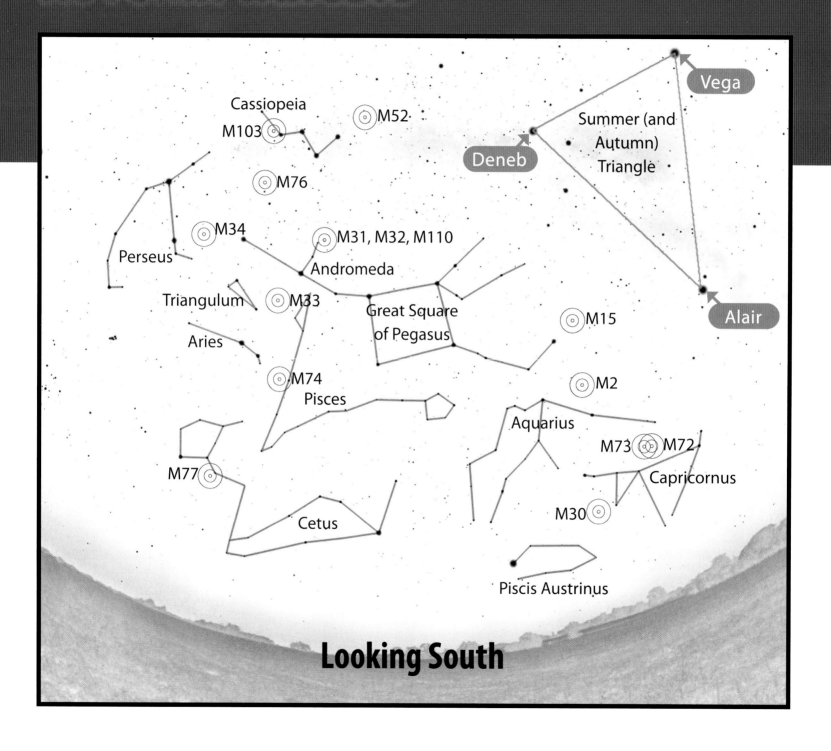

Looking South

Stargazing in Autumn

Autumn is one of the best times to stargaze. The Sun sets earlier, and the weather is generally pleasant. The lengthening nights bring with them an additional bonus. Many of the summer targets remain visible until the end of December!

Though the autumn sky is dominated by the constellation Pegasus, known by its "Great Square," this constellation contains rather few Messier objects. In fact, there is a swath of sky between the Andromeda Galaxy and globular cluster M15 that contains no Messier objects at all! When the Sun moves into that band during March every year, you can "run" your Messier Marathon.

We start our tour of the Messier list in autumn for a special reason. For those doing a Messier Marathon during the month of March, several of these targets can be observed shortly after sunset before they sink below the horizon. Note that the following pages present our suggested viewing order on an autumn evening.

The most popular autumn targets include the Andromeda Galaxy (M31), the Triangulum Galaxy (M33), and two impressive globular clusters, M15 and M2. Autumn also contains one of the most challenging Messier objects, M30, which is often missed by those doing a Messier Marathon in March, because it's easily swallowed up by the predawn light.

In September, Galaxy M74 rises around 9 p.m. and Galaxy M77 rises two hours later, at around 11 p.m. Since galaxies are more visible when they are higher in the sky, view these two Messier objects last on an autumn evening. In March, both of these galaxies set together, immediately after sunset, making them the very first two objects to observe during a Messier Marathon.

For Marathoners: Observing autumn targets in March is a bit more complicated because these targets are split between the evening and morning. For the winter, spring, and summer objects, we suggest that marathoners observe the objects in the order we list them. For those doing the Messier Marathon in March, here is our recommended observation order for the autumn targets:

After Sunset:	Before Dawn:
M74	M15
M77	M72
M33	M73
M31, M32, M110	M2
M52	M30 (very challenging)
M103	
M76	
M34	

M2

M2 is one of the largest globular clusters in our Galaxy's Halo. The largest globular cluster is Omega Centauri, which probably would have made Messier's list had he known about it. Omega Centauri, a southern sky target, was not visible from Messier's home in Paris.

Messier's telescope was not powerful enough to resolve the individual stars in this cluster. Instead, he referred to M2 as a "nebula without a star." It's interesting that this cluster hasn't had any popular names assigned to it. Perhaps we should simply call it the "Great Nameless Cluster in Aquarius."

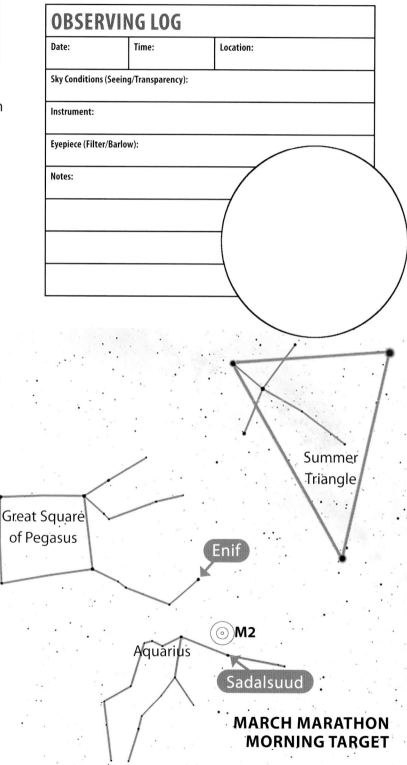

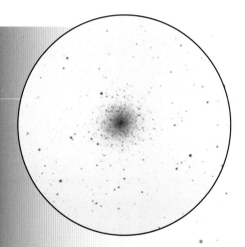

Common Name(s): Great Nameless Cluster in Aquarius
Type: Globular Cluster
Brightness (Visual Magnitude): 6.3
Distance (Light-Years): 39,000
Difficulty (Subjective): 2

Observing Tips: M2 is found in a rather empty part of the sky along the line connecting stars Enif (the Head of Pegasus) and Sadalsuud (the brightest star in Aquarius). High magnification (with large aperture) will provide stunning views.

MARCH MARATHON MORNING TARGET

M30

For many, observing M30 is one of the more challenging "suburb visible" targets in this book. It's low on the horizon (from northern latitudes), and if Fomalhaut is hidden by nearby houses, barely any other nearby stars are bright enough to see from within city limits.

If you're in dark skies with a clear view of the southern horizon, you should have little trouble finding M30, especially using the bright star Fomalhaut as a reference. This is also the hardest target for marathoners. In March, there are mere minutes between the time M30 rises above the horizon and sunrise.

Common Name(s): Jellyfish Cluster
Type: Globular Cluster
Brightness (Visual Magnitude): 7.3
Distance (Light-Years): 27,000
Difficulty (Subjective): 3–5

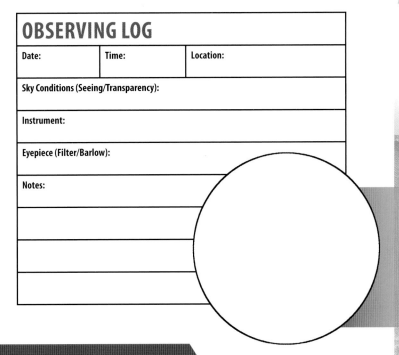

OBSERVING LOG

Date:	Time:	Location:

Sky Conditions (Seeing/Transparency):

Instrument:

Eyepiece (Filter/Barlow):

Notes:

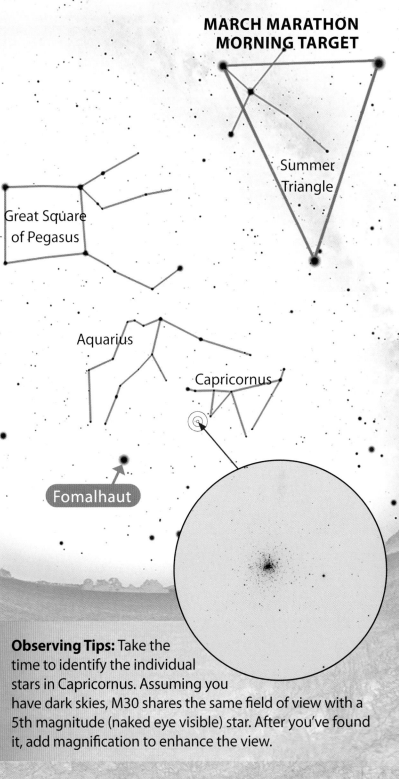

MARCH MARATHON MORNING TARGET

Summer Triangle

Great Square of Pegasus

Aquarius

Capricornus

Fomalhaut

Observing Tips: Take the time to identify the individual stars in Capricornus. Assuming you have dark skies, M30 shares the same field of view with a 5th magnitude (naked eye visible) star. After you've found it, add magnification to enhance the view.

M72

M72 is the most challenging globular cluster in Messier's list, and without many bright stars to use a reference, it can be a challenge to find even from dark skies. An 8-inch telescope or larger is recommended to resolve the stars in this cluster owing to its enormous distance of 54,500 light-years.

Without a large telescope, this cluster will look like a tiny smudge, like a spot of gray paint on a black canvas. At about 30x magnification, you can fit both M72 and M73 in the same field of view, although both objects will appear quite small.

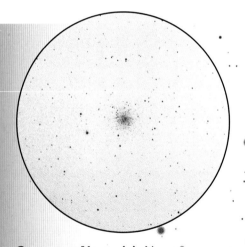

Common Name(s): None
Type: Globular Cluster
Brightness (Visual Magnitude): 9.3
Distance (Light-Years): 54,500
Difficulty (Subjective): 4

Observing Tips: One way to find M72 is to follow the Vega-Altair side of the Summer Triangle down to the two westernmost stars in Capricornus. These stars can be used to form a triangle with the westernmost side of Aquarius, with M72 found nearby.

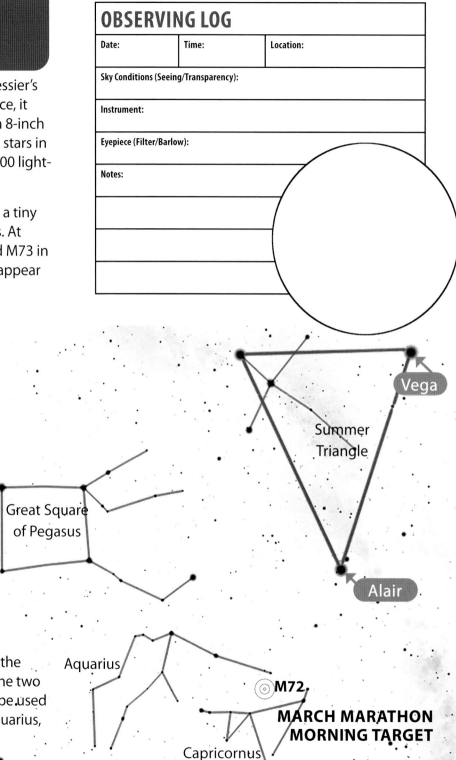

MARCH MARATHON
MORNING TARGET

M73

Well, if there was an award for the least interesting Messier object, M73 would probably win. This grouping of stars is simply a chance alignment of stars. In 2001, astrophysicists Michael Odenkirchen and Caroline Soubiran measured the distances to these stars and found that they range from under 1,000 light-years to over 3,000 light-years.

Our sky is covered in thousands of such alignments, so why is this a Messier object? Messier noted some nebulosity surrounding these stars (which is why he noted it in the first place); however, this was most likely due to either poor optics or poor seeing conditions.

Common Name(s): None
Type: Asterism
Brightness (Visual Magnitude): 8.9
Distance (Light-Years): 2,500
Difficulty (Subjective): 4

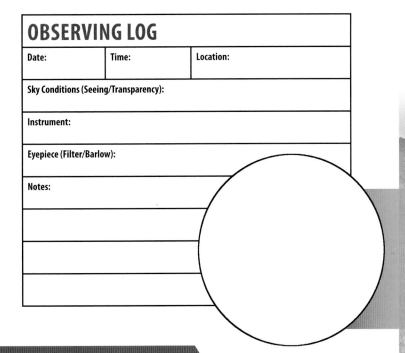

OBSERVING LOG

Date:	Time:	Location:

Sky Conditions (Seeing/Transparency):

Instrument:

Eyepiece (Filter/Barlow):

Notes:

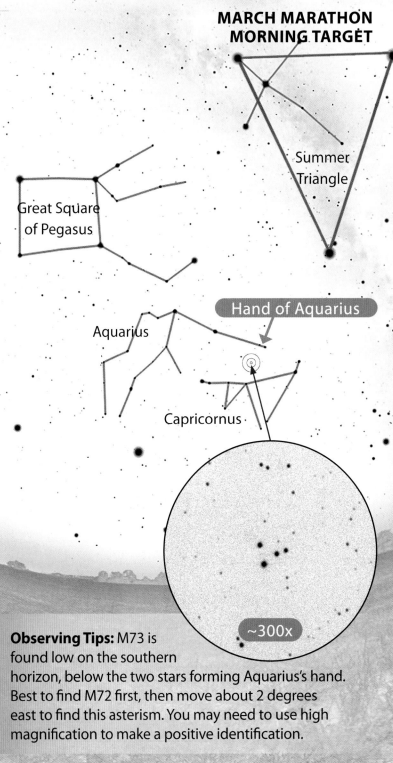

Observing Tips: M73 is found low on the southern horizon, below the two stars forming Aquarius's hand. Best to find M72 first, then move about 2 degrees east to find this asterism. You may need to use high magnification to make a positive identification.

M15

M15 is one of the brightest globular clusters in the northern sky and is a great target to share at stargazing events (along with its neighbor M2). M15 is recognizable, as it sits within a triangle of three prominent foreground stars.

M15 is relatively easy to find in the autumn. The challenge with finding M15 during a Marathon (in March) is that the Great Square is partially below the horizon, so you'll need to use the Northern Cross as reference. One you've identified the star Enif, extend the horse's neck toward Vega. M15 is just one bullseye width away.

OBSERVING LOG

Date:	Time:	Location:

Sky Conditions (Seeing/Transparency):

Instrument:

Eyepiece (Filter/Barlow):

Notes:

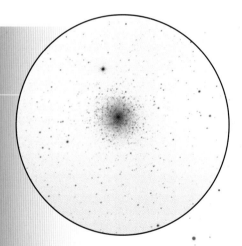

Common Name(s): Pegasus Cluster
Type: Globular Cluster
Brightness (Visual Magnitude): 6.0
Distance (Light-Years): 32,600
Difficulty (Subjective): 2

Observing Tips: Once you've found M15, switch to higher powered eyepieces and use averted vision to increase contrast and observe the dimmer stars in the cluster. Globular clusters really pop in larger aperture telescopes where hundreds of individual stars are clearly visible.

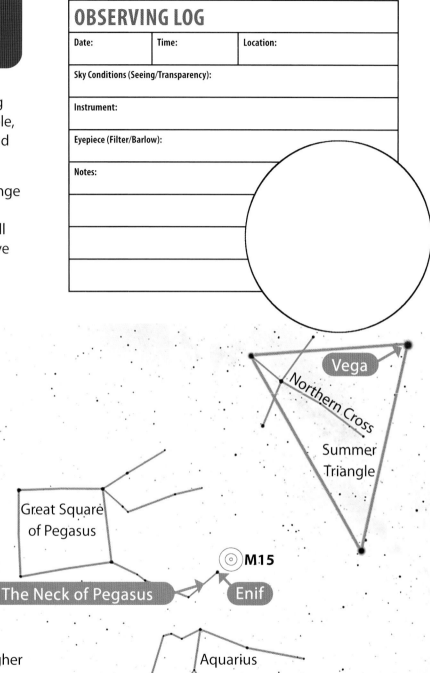

MARCH MARATHON
MORNING TARGET

M52

M52 is a dense open cluster containing over 200 stars. Though most of the stars appear white, keep an eye out for red, orange, and blue stars nearby. These brighter colored stars are in the foreground and are not members of the cluster itself.

Astrophotographers are quite familiar with this cluster's location, as it lies right beside the Bubble Nebulae, a target easily picked up by cameras, but a challenge for the eye.

Common Name(s): Salt-and-Pepper Cluster
Type: Open Cluster
Brightness (Visual Magnitude): 6.9
Distance (Light-Years): 5,000
Difficulty (Subjective): 2

OBSERVING LOG

Date:	Time:	Location:

Sky Conditions (Seeing/Transparency):

Instrument:

Eyepiece (Filter/Barlow):

Notes:

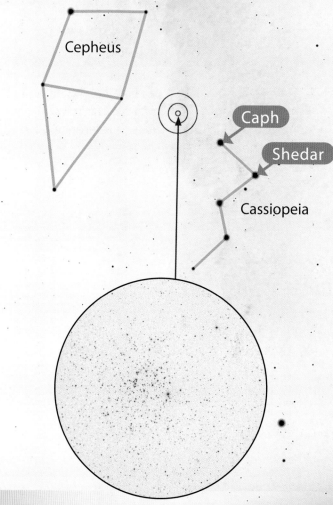

Observing Tips: To locate M52, double the line connecting Shedar to Caph, the two end stars of Cassiopeia. When observing M52 as part of a March Marathon, the cluster will be located below Cassiopeia, just above the horizon in the northwest.

M76

Despite its location between Cassiopeia and Andromeda, M76 is technically part of the Perseus constellation (we've shown the constellation boundaries on this map).

M76 is arguably the most intricate of the Messier planetary nebulae, although it is much smaller than M27 (the Dumbbell Nebula) and far less famous than M57 (the Ring Nebula). In dark skies, M76 will first appear as a stubby bar, but increased magnification (to improve contrast) may help to reveal its two gaseous lobes.

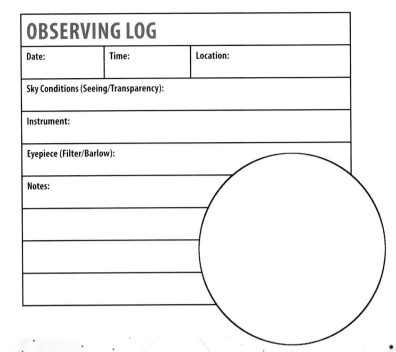

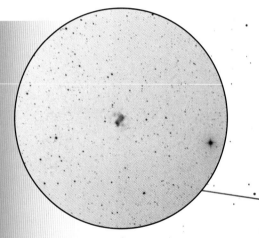

Common Name(s): Little Dumbbell Nebula
Type: Planetary Nebula
Brightness (Visual Magnitude): 10.1
Distance (Light-Years): 3,400
Difficulty (Subjective): 5

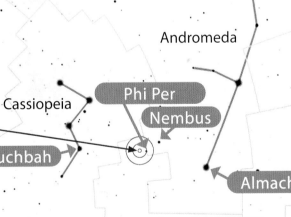

Observing Tips: Look for M76 directly between Ruchbah in Cassiopeia and Almach in Andromeda, close to the medium-bright stars Phi Per and Nembus (a.k.a. 51 And). For best results, use a narrowband filter. Be sure to experiment with different eyepieces. Depending on your seeing conditions and light pollution level, the view may (or may not) improve with magnification.

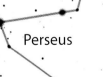

Autumn

M103

M103 is a beautiful, bright star cluster located very close to the star Ruchbah in Cassiopeia. The cluster is bisected by three bright stars, the central one a red giant. It reminds Chris of the Jewel Box Cluster in the constellation Crux, a favorite he observed during a trip to Australia.

It's interesting that early accounts of this cluster record it as a nebula. This is most likely due to the limitations of early telescopes. M103 was the last entry in Messier's original list published in the 1784 issue of *Connaissance des temps*.

Common Name(s): Traffic Light Cluster
Type: Open Cluster
Brightness (Visual Magnitude): 7.4
Distance (Light-Years): 9,800
Difficulty (Subjective): 2

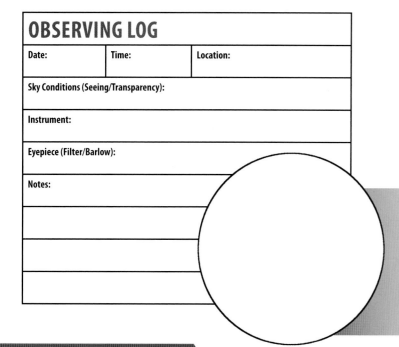

OBSERVING LOG

Date:	Time:	Location:

Sky Conditions (Seeing/Transparency):

Instrument:

Eyepiece (Filter/Barlow):

Notes:

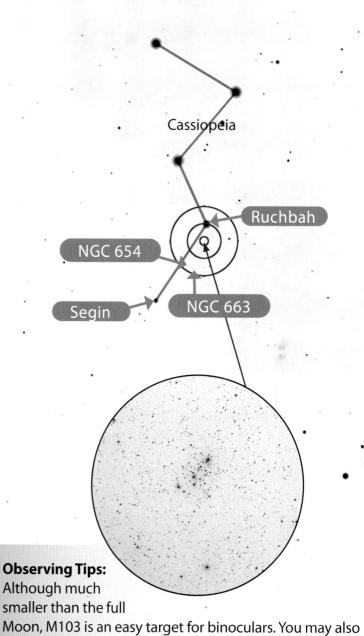

Cassiopeia

Ruchbah

NGC 654

Segin

NGC 663

Observing Tips:
Although much smaller than the full Moon, M103 is an easy target for binoculars. You may also want to observe neighboring star clusters NGC 654 and NGC 663, also located between Ruchbah and Segin.

M31, M32, M110

M31 is our Milky Way Galaxy's big brother. It's about twice as large and similar in form. Located only 2.7 million light-years away, we're basically neighbors! In a few billion years, M31 and our galaxy will collide, eventually forming an even larger galaxy after a billion-year-long cosmic dance.

M32 and M110 are dwarf, satellite galaxies to M31. M32 is closer than M31, and M110 is slightly farther. M110 was the final object added to Messier's original list of 103 objects published in 1781.

OBSERVING LOG

Date:	Time:	Location:

Sky Conditions (Seeing/Transparency):

Instrument:

Eyepiece (Filter/Barlow):

Notes:

Common Name(s): Andromeda Galaxy
Type: Spiral Galaxy
Brightness (Visual Magnitude): 3.4 (M31), 8 (M32 & M110)
Distance (Light-Years): 2.7 million
Difficulty (Subjective): 1 (M31), 2 (M32 & M110)

Observing Tips: M31 can be seen without a telescope from dark skies. Its size (6 times the width of the full Moon) makes it ideal for binoculars. Use a low magnification to see details in M31's disk and its two companion galaxies.

Need a quick way to find this galaxy? The deeper "V" in the "W" of Cassiopeia points (approximately) toward M31.

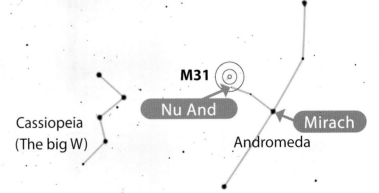

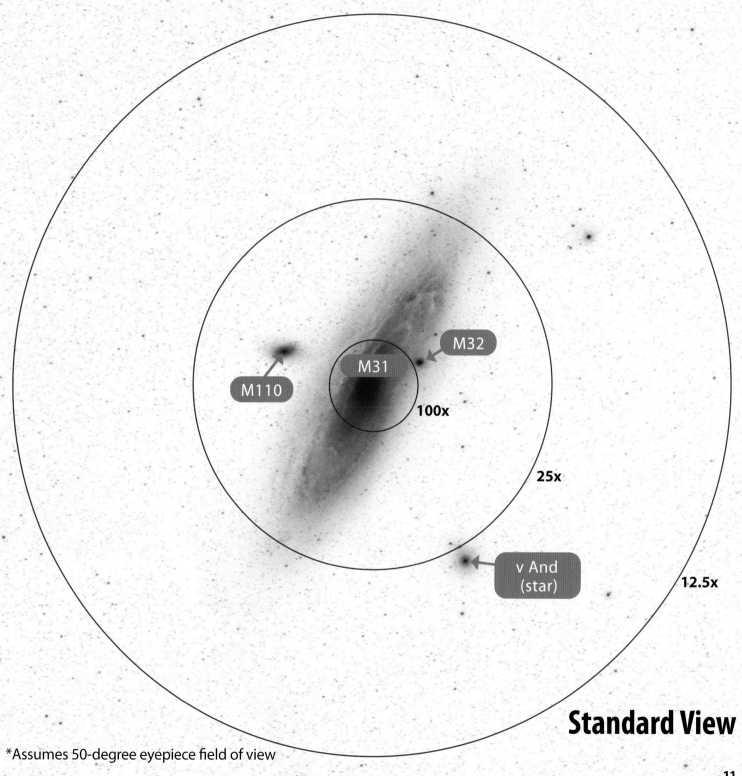

M110

M31

M32

100x

25x

v And
(star)

12.5x

Standard View

*Assumes 50-degree eyepiece field of view

11

M33

M33 is sort of like the Andromeda Galaxy's little cousin. It's at a similar distance but is smaller and much less massive. M33 is close enough to us that large amateur telescopes can even observe star clusters and faint nebula within the galaxy!

This galaxy has a very low surface brightness, which makes it blend into light polluted skies. I was able to find it, with some difficulty, using an 8-inch telescope from the suburbs of San Francisco. Some people have claimed to see M33 with their unaided eyes under very dark skies.

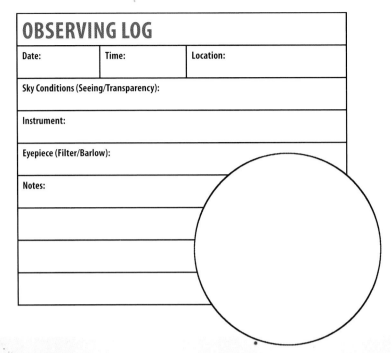

OBSERVING LOG

Date:	Time:	Location:

Sky Conditions (Seeing/Transparency):

Instrument:

Eyepiece (Filter/Barlow):

Notes:

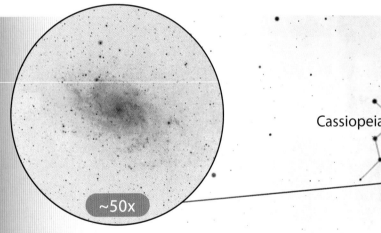

~50x

Common Name(s): Triangulum Galaxy
Type: Spiral Galaxy
Brightness (Visual Magnitude): 5.7
Distance (Light-Years): 2.7 million
Difficulty (Subjective): 3 (due to very low surface brightness)

Observing Tips: To find M33, use the deeper "V" of Cassiopeia's big "W" as an arrow to hop over to the Andromeda constellation. Follow an imaginary line from the star "Mu And" through Mirach, about midway to Hamal/Sheratan. In mildly light polluted skies, try increasing the magnification to improve the contrast in the spiral arms.

M34

M34 is one of three large, bright, open clusters in Perseus. The other two, a close-together pair known as the Double Cluster (NGC 884 and NGC 869), were probably too bright to be included in Charles Messier's list of "things that are not comets."

When observing with new stargazers and teaching them to use telescopes, M34 is often one of our first targets. It's easy to find and looks beautiful at low magnification from a small city. M34 is found midway between Almach, in Andromeda, and Algol (the Demon Star), an eclipsing binary star that dims significantly every three days.

Common Name(s): Spiral Cluster
Type: Open Cluster
Brightness (Visual Magnitude): 5.2
Distance (Light-Years): 1,500
Difficulty (Subjective): 1

OBSERVING LOG

Date:	Time:	Location:
Sky Conditions (Seeing/Transparency):		
Instrument:		
Eyepiece (Filter/Barlow):		
Notes:		

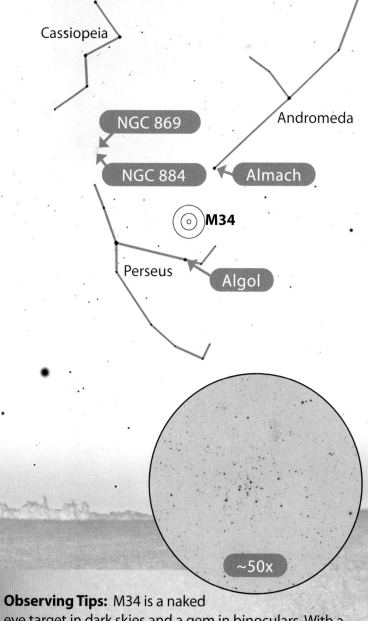

Cassiopeia

NGC 869

NGC 884

Andromeda

Almach

M34

Perseus

Algol

~50x

Observing Tips: M34 is a naked eye target in dark skies and a gem in binoculars. With a telescope, be sure to use your lowest magnification or widest angle eyepiece.

M74

This face-on spiral galaxy has the nickname the "Phantom" because it is nearly invisible in all but the darkest skies. Cameras easily pick up the galaxy's spiral structure. In the eyepiece, you should see a faint halo around a brighter core. Several famous astronomers originally mistook M74 for a distant globular cluster.

M74 is best viewed in late evening during autumn. If you're doing a mid-March Messier Marathon, M74 and M77 should be your first targets. Find the galaxies above the western horizon as soon as the sky darkens—before they drop too low to see clearly.

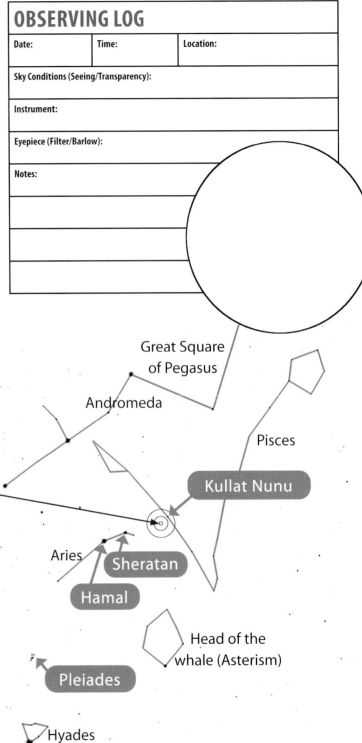

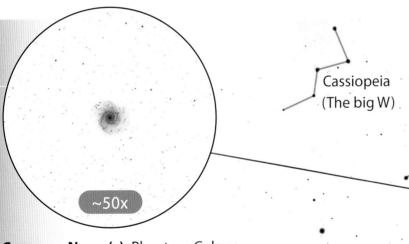

~50x

Common Name(s): Phantom Galaxy
Type: Spiral Galaxy
Brightness (Visual Magnitude): 9.4
Distance (Light-Years): 30 million
Difficulty (Subjective): 5

Observing Tips: Use the two bright stars in Aries to guide you to M74, found near the star Kullat Nunu in Pisces. In March, with Cassiopeia lower in the sky and the Great Square on the horizon, you may want to turn this map on its side and hop down to Aries from the Hyades.

Cassiopeia (The big W)

Great Square of Pegasus

Andromeda

Pisces

Kullat Nunu

Aries

Sheratan

Hamal

Head of the whale (Asterism)

Pleiades

Hyades

M77

M77 is slightly easier to find than M74. At first it may look like a double star, since M77 is located behind a similarly bright foreground star. M77 has a dense core that isn't too difficult to identify, but resolving details in this galaxy's arms will require very dark skies, and quite a bit of patience.

Fortunately, M77 is located in a relatively accessible location. Starting at the bright star Menkar, identify the fainter stars marking the "neck" of Cetus, a constellation that represents a sea creature in Greek mythology.

Common Name(s): Cetus A
Type: Spiral Galaxy
Brightness (Visual Magnitude): 8.9
Distance (Light-Years): 47 million
Difficulty (Subjective): 4

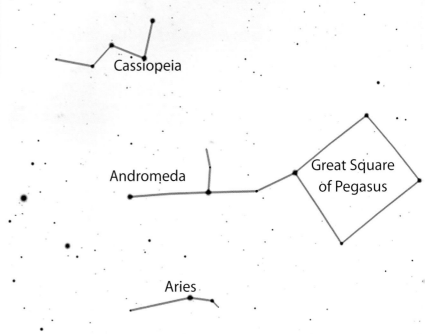

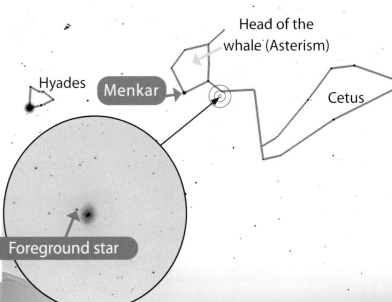

Foreground star

Observing Tips: Use a higher power eyepiece to increase contrast and to observe this galaxy's spiral structure. During a mid-March Marathon you'll only have a few minutes to view M77 and M74 before they sink below the horizon.

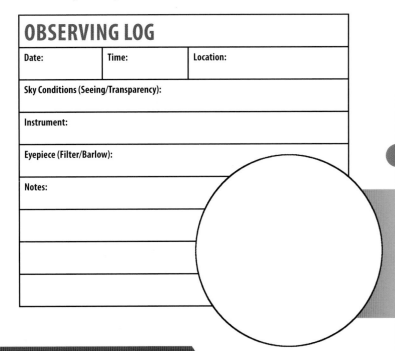

OBSERVING LOG

Date:	Time:	Location:

Sky Conditions (Seeing/Transparency):

Instrument:

Eyepiece (Filter/Barlow):

Notes:

WINTER TARGETS

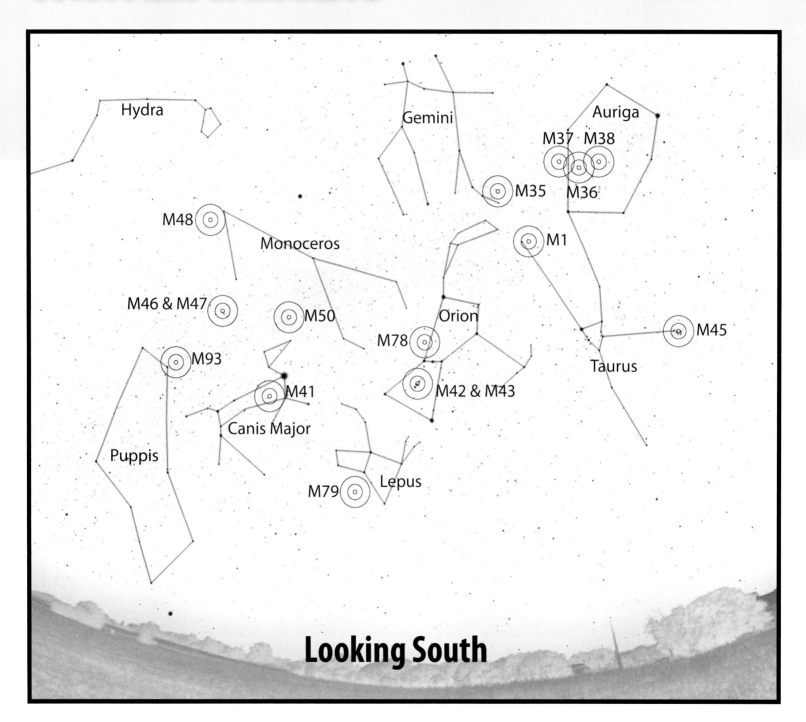

Hydra

Gemini

Auriga

M37 M38

M35

M36

M48

M1

Monoceros

M46 & M47

M50

Orion

M45

M78

Taurus

M93

M42 & M43

M41

Canis Major

Puppis

Lepus

M79

Looking South

Stargazing in Winter

It can be a challenge to motivate yourself to stargaze during the winter months, but if you do, you'll be rewarded with some of our sky's most treasured targets.

Each December, social media is flooded with images of the Great Orion Nebula as budding astrophotographers focus their cameras on one of the most photographed objects in the sky. The Orion Nebula (M42) is one of the few nebulae easily visible from the city, and it doesn't disappoint.

Eleven of the of the fifteen winter objects are open star clusters. Most of these are visible from the suburbs, or even the city without too much difficulty. M45, better known as the Pleiades, is bright enough to see with unaided eyes, even from the city. This open cluster is spectacular with the added magnification provided by binoculars and telescopes.

M79 is the only globular cluster this season, and there are no galaxies in this list of winter targets at all! That doesn't mean there aren't any galaxies to see at this time of year. Spring's M81 and M82 are circumpolar for much of the Northern Hemisphere and make great galaxy targets during winter. M31, the Andromeda Galaxy, is high above the horizon for early evening viewing.

The most challenging object to see from the city is M78 (a reflection nebula). M1, the Crab Nebula supernova remnant, is also a challenge from the suburbs, but not impossible. These nebulae have low surface brightness and tend to get lost in light polluted skies.

If you're working on the Messier Marathon, the viewing order doesn't matter too much here. All of these objects are found in the southern sky just after sunset and will take several hours to fall below the horizon. You'll want to make sure you observe M79 early on though, or you'll risk losing it behind a tree.

Looking for more targets? Be sure to observe the colored double star Winter Albiero (found on the M93 page of this book) and the Hyades, a nearby open cluster, found alongside M45.

Between ice storms and the occasional nor'easter, the top of a picnic table is often the only clear place to set John's 12-inch telescope (the dog's name is Lyra).

M45

The Pleiades, M45, is probably the best known Messier object. This tight cluster of bright stars has been known since antiquity and is found in the mythologies of many cultures. You may often hear people call it the 7 Sisters, or Subaru.

The Pleiades features prominently in the winter sky within the constellation Taurus and is visible even under light polluted skies. Since the Pleiades lies near the ecliptic, it is sometimes found near the Moon or planets. On dark transparent nights, Chris has been able to see nebulosity around the main stars in his 8-inch telescope.

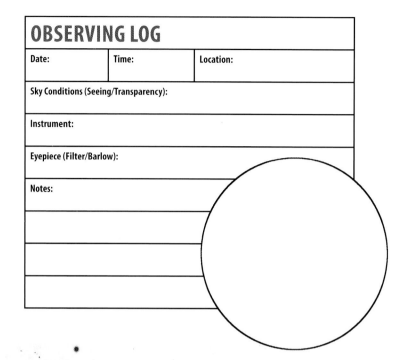

OBSERVING LOG

Date:	Time:	Location:

Sky Conditions (Seeing/Transparency):

Instrument:

Eyepiece (Filter/Barlow):

Notes:

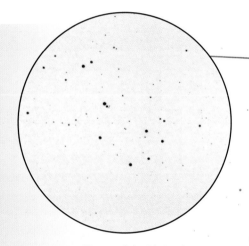

Common Name(s): Pleiades
Type: Open Cluster
Brightness (Visual Magnitude): 1.2
Distance (Light-Years): 380
Difficulty (Subjective): 1

Observing Tips: Beginner stargazers often confuse M45 with the Little Dipper. To the naked eye, only 6 or 7 stars are visible, but binoculars reveal many additional stars within the cluster. To view this cluster in its entirety with your telescope, use your lowest power eyepiece.

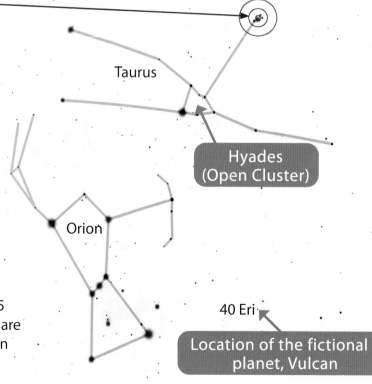

Taurus

Hyades
(Open Cluster)

Orion

40 Eri

Location of the fictional
planet, Vulcan

18

M79

M79 is one of the few globular clusters visible in winter. With a declination of 24 degrees south, this cluster appears low in the winter sky. From here at about 45 degrees north, it barely rises 20 degrees over our horizon.

M79 is too low for the domed telescope at my University to reach. To view it, I'll take a Dobsonian telescope out onto our school's observation deck where we have a southern facing view over the Atlantic Ocean.

Common Name(s): None
Type: Globular Cluster
Brightness (Visual Magnitude): 7.8
Distance (Light-Years): 42,100
Difficulty (Subjective): 3

OBSERVING LOG

Date:	Time:	Location:
Sky Conditions (Seeing/Transparency):		
Instrument:		
Eyepiece (Filter/Barlow):		
Notes:		

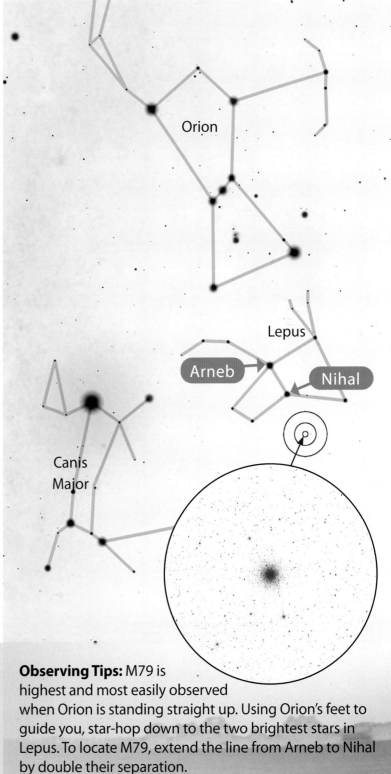

Observing Tips: M79 is highest and most easily observed when Orion is standing straight up. Using Orion's feet to guide you, star-hop down to the two brightest stars in Lepus. To locate M79, extend the line from Arneb to Nihal by double their separation.

M42 & M43

M42 is bright enough that unaided eyes will see it as the middle "star" in Orion's sword, hanging just below his belt. In your telescope, the Orion Nebula will show complex veils and a central clump of stars named "The Trapezium" because of their trapezoidal arrangement.

M43 is a smaller and fainter section of the same gas cloud. It is separated from M42 by a dark dust lane. M43 is quite intricate and worthy of its own designation. Look for its central star, a variable double star named NU Orionis, or "Bonds Star."

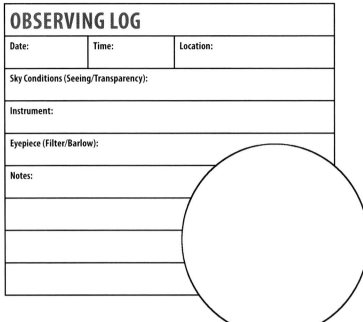

OBSERVING LOG

Date:	Time:	Location:

Sky Conditions (Seeing/Transparency):

Instrument:

Eyepiece (Filter/Barlow):

Notes:

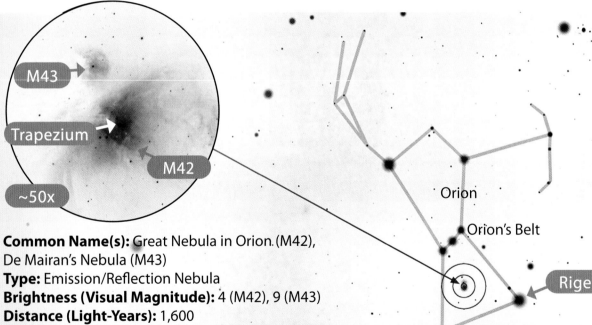

M43

Trapezium

M42

~50x

Orion

Orion's Belt

Rigel

Common Name(s): Great Nebula in Orion (M42), De Mairan's Nebula (M43)
Type: Emission/Reflection Nebula
Brightness (Visual Magnitude): 4 (M42), 9 (M43)
Distance (Light-Years): 1,600
Difficulty (Subjective): 1

Observing Tips: The Orion Nebula looks great at all magnifications and will be enhanced by a narrowband filter. Be sure to experiment with different eyepieces to see which one gives you the most pleasing views. M42 and M43 are also great targets for binoculars in almost any sky conditions.

M78

M78 is quite dim and best reserved for dark sky observation (far from city lights and on moonless nights). When you resolve this nebula in your telescope, there are two distinct sections. The bulk of this nebula, designated M78, is complemented by a second, although dimmer, portion, NGC 2071. M78 reminds Chris of a comet that is splitting apart.

These nebulae are part of the same general gaseous region as the Orion Nebula. Larger telescopes and extremely dark skies will reveal several other nebulae within the Orion constellation.

Common Name(s): Casper the Friendly Ghost
Type: Reflection Nebula
Brightness (Visual Magnitude): 8.3
Distance (Light-Years): 1,600
Difficulty (Subjective): 4

OBSERVING LOG

Date:	Time:	Location:

Sky Conditions (Seeing/Transparency):

Instrument:

Eyepiece (Filter/Barlow):

Notes:

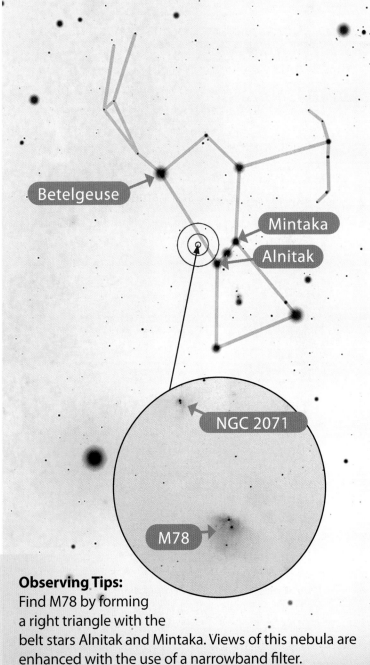

Betelgeuse

Mintaka

Alnitak

NGC 2071

M78

Observing Tips:
Find M78 by forming a right triangle with the belt stars Alnitak and Mintaka. Views of this nebula are enhanced with the use of a narrowband filter.

M1

In the year 1054, Chinese astronomers observed a bright object in the sky, which we now know was a supernova, an exploding star. The explosion of gas and dust, which is still expanding to this day, was rediscovered in 1731 by an English astronomer named John Bevis.

Charles Messier observed the object in 1758, assigning it to be the first item in his catalog of "Objects that are not comets." To find M1, use Orion's Belt or the Hyades star cluster to identify the tip of the lower horn in Taurus (the Bull). At low magnification, M1 is found in the same field of view as the star known as Tianguan (a.k.a. Zeta Tau).

OBSERVING LOG

Date:	Time:	Location:

Sky Conditions (Seeing/Transparency):

Instrument:

Eyepiece (Filter/Barlow):

Notes:

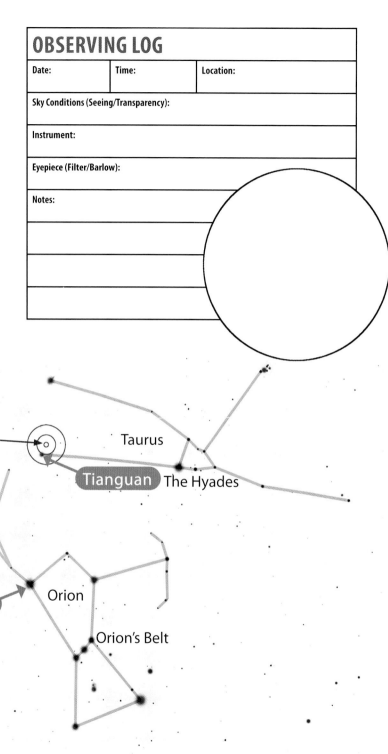

Common Name(s): Crab Nebula
Type: Supernova Remnant
Brightness (Visual Magnitude): 8.4
Distance (Light-Years): 6,300
Difficulty (Subjective): 4

Observing Tips: M1 is smaller than you might expect, and while the Orion Nebula is visible in most seeing conditions, M1 is much more challenging, with no obvious internal stars. Try to view it when it is high in the sky. A narrowband filter will help.

22

M37

The many prominent stars in this cool-looking open cluster suggest shapes like a boomerang or a shimmering Romulan Warbird from *Star Trek* beginning to cloak.

M37 is the brightest in a line of three open clusters that cross the middle of the constellation Auriga. It can just barely be seen as a tiny patch with the unaided eye in extremely dark skies, but is quite easily visible in binoculars. M37 is a relatively easy target to pick up in any telescope, even from a small city.

Common Name(s): Warbird Cluster, Auriga Salt-and-Pepper Cluster
Type: Open Cluster
Brightness (Visual Magnitude): 5.6
Distance (Light-Years): 4,400
Difficulty (Subjective): 2

OBSERVING LOG

Date:	Time:	Location:

Sky Conditions (Seeing/Transparency):

Instrument:

Eyepiece (Filter/Barlow):

Notes:

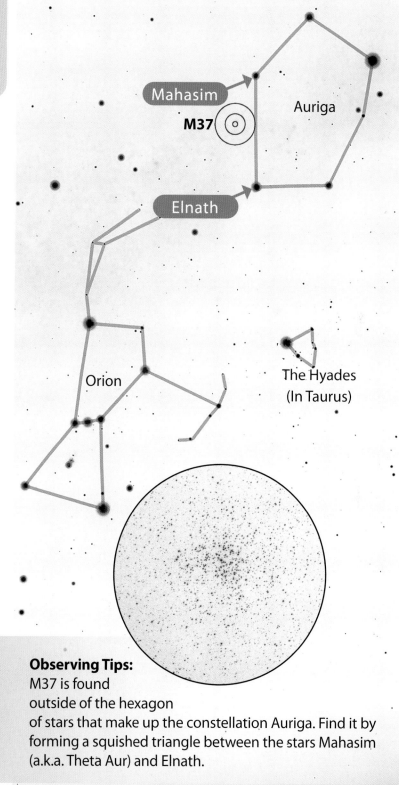

Observing Tips:
M37 is found outside of the hexagon of stars that make up the constellation Auriga. Find it by forming a squished triangle between the stars Mahasim (a.k.a. Theta Aur) and Elnath.

M36

The constellation Auriga has several easy-to-find open clusters (including three Messier objects), but M36 is John's personal favorite. Even when sharing his telescope's view with the public, everyone can see this cluster's structure with ease.

John often puts M36 in one of the telescopes on his school's observation deck and asks students what they think it looks like. That's where the name "the Turtle Cluster" originated, and now he has trouble imagining it as anything else.

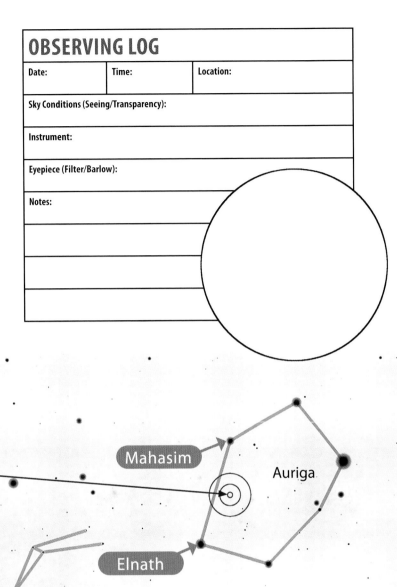

OBSERVING LOG

Date:	Time:	Location:

Sky Conditions (Seeing/Transparency):

Instrument:

Eyepiece (Filter/Barlow):

Notes:

Common Name(s): The Pinwheel Cluster, The Turtle Cluster
Type: Open Cluster
Brightness (Visual Magnitude): 6.0
Distance (Light-Years): 4,100
Difficulty (Subjective): 2

Observing Tips: M36 is one of the easiest clusters to find, even with moderate magnification. To find it, form a squashed triangle inside Auriga between Elnath and Mahasim.

24

M38

This open cluster's main stars form a prominent "X" shape—but that pattern may become obscured under very dark skies, where more background stars become visible (as is the case in our "eyepiece view" image below). While you're in the area, see if you can locate open cluster NGC 1907; at low magnification, it may be in the same field of view as M38.

Is the Moon in the sky? A lunar feature nicknamed the Lunar X sometimes appears on the Moon near first quarter. Check the Internet to confirm exactly when the Lunar X will be visible.

Common Name(s): Starfish Cluster, Cluster X
Type: Open Cluster
Brightness (Visual Magnitude): 6.4
Distance (Light-Years): 4,200
Difficulty (Subjective): 2 or 3

OBSERVING LOG

Date:	Time:	Location:
Sky Conditions (Seeing/Transparency):		
Instrument:		
Eyepiece (Filter/Barlow):		
Notes:		

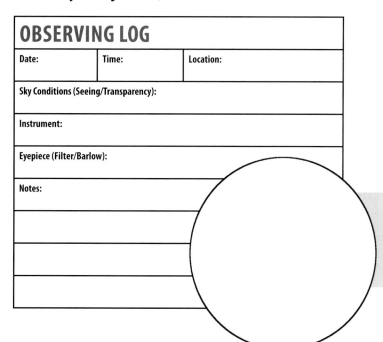

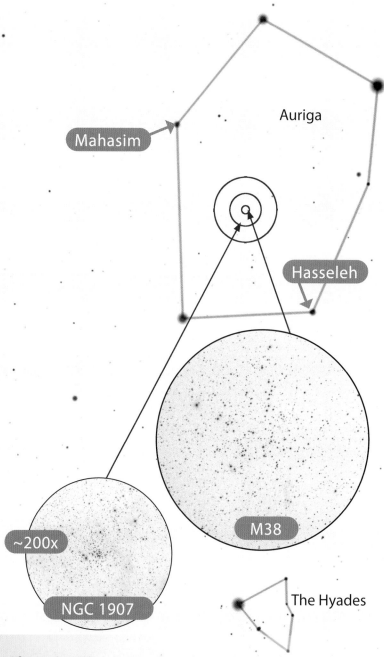

Mahasim

Auriga

Hasseleh

~200x

NGC 1907

M38

The Hyades

Observing Tips: M38 is found near the center of the hexagon of stars that make up the constellation Auriga, midway between Mahasim and Hasseleh.

M35

M35 is fairly large, about the size of the full Moon. This open cluster is visible in binoculars, even in moderately light polluted skies. From a dark sky location, M35 can be seen with the unaided eye, appearing as a patch of loosely scattered bright stars.

With a telescope, another open cluster named NGC 2158 is visible in the background. Although NGC 2158 appears smaller than M35, it is roughly the same size, occupying a similar volume of space within our galaxy.

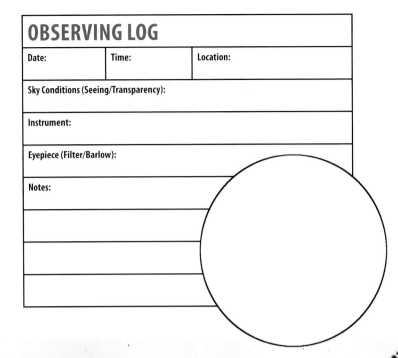

OBSERVING LOG

Date:	Time:	Location:

Sky Conditions (Seeing/Transparency):

Instrument:

Eyepiece (Filter/Barlow):

Notes:

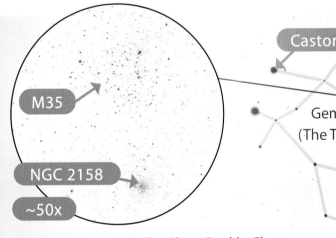

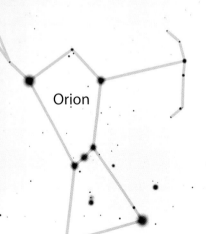

Common Name(s): The Shoe-Buckle Cluster
Type: Open Cluster
Brightness (Visual Magnitude): 5.1
Distance (Light-Years): 2,800
Difficulty (Subjective): 2

Observing Tips: To find M35, line up the four feet in the Gemini Twins. The cluster is found near the toe of the right-most twin (Castor).

M41

M41 is found inside the body of the "greater dog," Canis Major. The cluster appears slightly larger than the full Moon and can be seen with binoculars under most conditions, and with the naked eye from dark skies.

M41 hosts several interesting colored stars in a telescope. Pay attention to a bright red star near its center, and a 6th magnitude blue star named 12 CMa sitting a short distance outside of the cluster's lower edge. The blue star is nearly invisible to the unaided eye, but bright and beautiful through a telescope.

Common Name(s): Little Beehive Cluster
Type: Open Cluster
Brightness (Visual Magnitude): 4.5
Distance (Light-Years): 2,300
Difficulty (Subjective): 2

OBSERVING LOG

Date:	Time:	Location:
Sky Conditions (Seeing/Transparency):		
Instrument:		
Eyepiece (Filter/Barlow):		
Notes:		

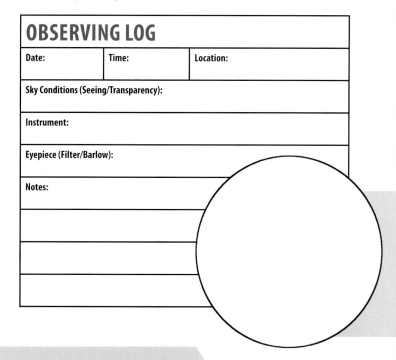

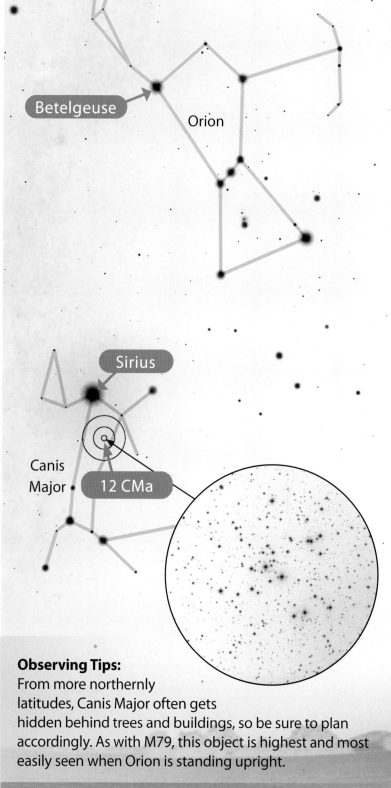

Betelgeuse

Orion

Sirius

Canis Major

12 CMa

Observing Tips:
From more northernly latitudes, Canis Major often gets hidden behind trees and buildings, so be sure to plan accordingly. As with M79, this object is highest and most easily seen when Orion is standing upright.

M93

M93 appears quite tiny and compact compared to the other open clusters we've seen so far this season. It can be a challenge to find due to lack of nearby reference stars. While technically within the constellation Puppis, you'll probably want to hop over from the brighter stars in Canis Major.

While you're in the area, be sure to check out one of the most popular colored double stars, 145 CMa, nicknamed Winter Albireo. To find Winter Albireo, form an equilateral triangle with the two lower stars in the dog's back.

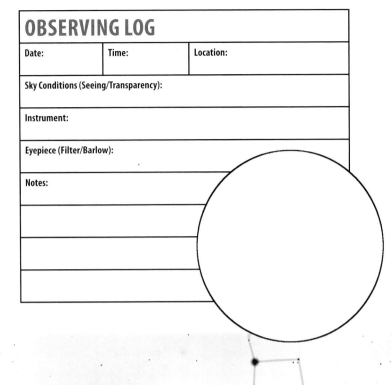

Orion

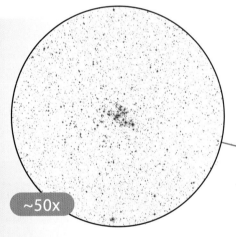

~50x

145 CMa (Double Star)

Azmidi (Double Star)

Canis Major (Greater Dog)

Puppis

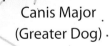

Common Name(s): Winter Butterfly Cluster
Type: Open Cluster
Brightness (Visual Magnitude): 6.2
Distance (Light-Years): 3,600
Difficulty (Subjective): 3

Observing Tips: To find M93, form a tall triangle with the stars in the greater dog's tail with the double star Azmidi (a.k.a. 7 Pup) in Puppis. At low magnification, the cluster and this double star fit into the same field of view.

M50

M50's home constellation of Monoceros is composed of faint stars that are a challenge to see in a light-polluted sky. Fortunately, this open cluster is quite bright and sits close to the more recognizable stars of Canis Major.

The "telescope view" of M50 shown below really doesn't do this cluster justice. Observed through a telescope at low power, its brighter stars truly resemble a heart. But that pattern can become obscured on dark nights when the background stars are enhanced.

Common Name(s): Heart-Shaped Cluster
Type: Open Cluster
Brightness (Visual Magnitude): 5.9
Distance (Light-Years): 3,000
Difficulty (Subjective): 3

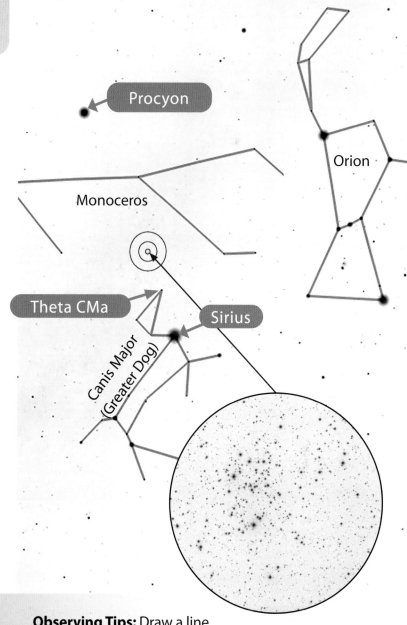

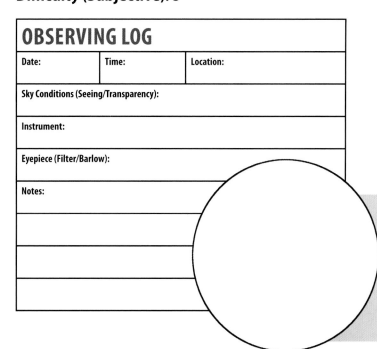

OBSERVING LOG

Date:	Time:	Location:
Sky Conditions (Seeing/Transparency):		
Instrument:		
Eyepiece (Filter/Barlow):		
Notes:		

Observing Tips: Draw a line from Sirius through the dog's nose (Theta CMa) and double its length. Alternatively, note that M50's distance from Sirius is the diameter of your fist held at arm's length, and simply search the imaginary line between Sirius and Procyon.

M46 & M47

M46 is the fainter of two bright open clusters located side-by-side near the northern boundary of the constellation Puppis. Both clusters are visible with unaided eyes in a dark sky. At low magnification, both M46 and M47 can fit into the same telescope field of view. If you have a very dark sky, increase the magnification and look for the planetary nebula NGC 2438 just offset from the core of M46.

M47 is the brighter companion to M46 (it's brighter because it's closer), though it appears to contain far fewer stars.

Common Name(s): Winter Double Cluster
Type: Open Cluster
Brightness (Visual Magnitude): 6.1 (M46), 4.4 (M47)
Distance (Light-Years): 5,400 (M46), 1,600 (M47)
Difficulty (Subjective): 2 (M47), 3 (M46)

Observing Tips: M46 and M47 can be found by tracing an imaginary line from Sirius through the star at the top of the dog's head (Muliphein).

As you can see from the image on the right, 25x magnification is sufficient for getting both M46 and M47 in the same field of view. However, to view the planetary nebula, NGC 2438, you may want to increase the magnification and enlarge M46 alone.

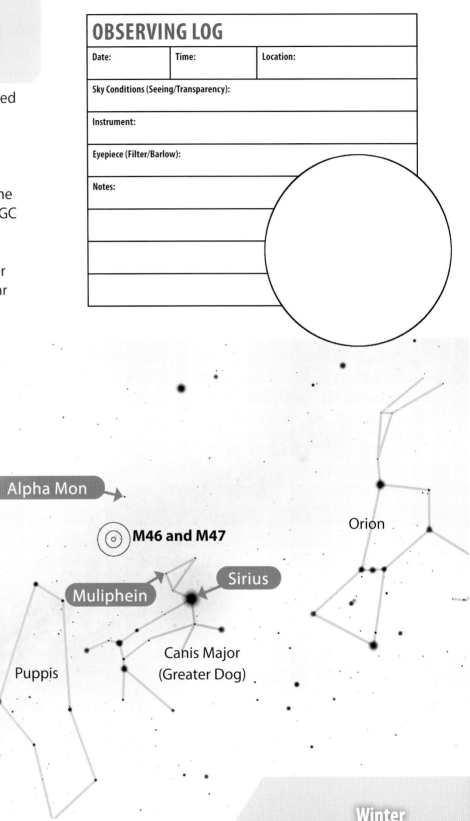

OBSERVING LOG		
Date:	Time:	Location:
Sky Conditions (Seeing/Transparency):		
Instrument:		
Eyepiece (Filter/Barlow):		
Notes:		

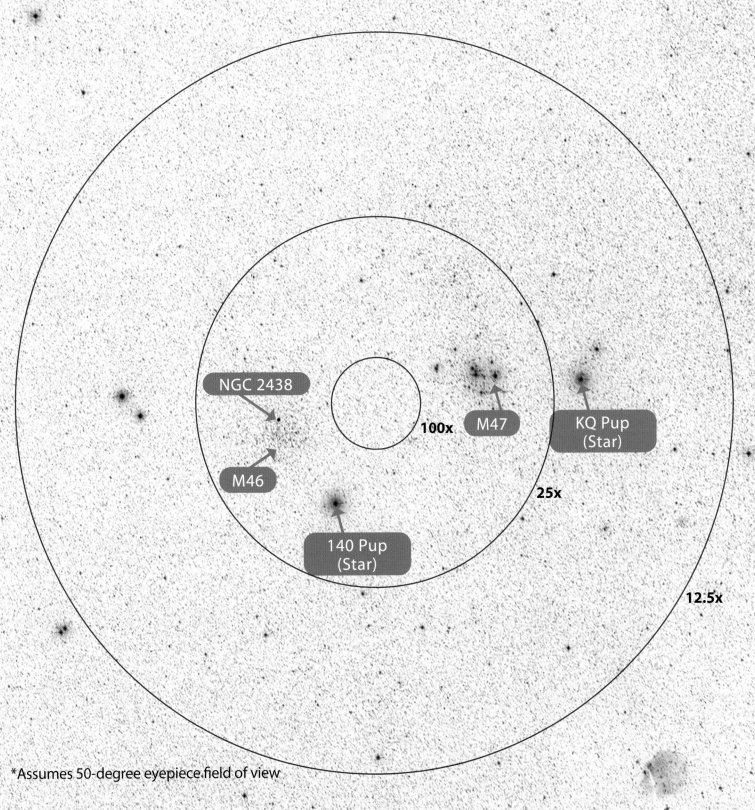

NGC 2438

M46

100x

M47

KQ Pup
(Star)

25x

140 Pup
(Star)

12.5x

*Assumes 50-degree eyepiece field of view

M48

M48 is referred to as Messier's "missing" object. That's because Messier's recording of its location is several degrees off. Author Steven O'Meara notes two "naked-eye glows" in this region, the dimmer of which is M48, the brighter one an "unresolved loose gathering of stars [through slightly unfocused binoculars]." O'Meara speculates that the latter may be the "real" M48.

Without nearby bright stars, M48 can be a bit challenging to find. Fortunately, the cluster itself is fairly bright. If you can identify C Hya and Alpha Mon, M48 is found along the imaginary line between them.

OBSERVING LOG

Date:	Time:	Location:
Sky Conditions (Seeing/Transparency):		
Instrument:		
Eyepiece (Filter/Barlow):		
Notes:		

1, 2 and C Hydrae

M48

~20x Magnification

Hydra

1, 2 and C Hydrae

Orion

Monoceros

Alpha Mon

Common Name(s): None
Type: Open Cluster
Brightness (Visual Magnitude): 5.8
Distance (Light-Years): 1,500
Difficulty (Subjective): 3

Observing Tips: While searching for M48, you'll probably notice a trio of bright stars. These are 1, 2, and C Hydrae. At around 20x magnification, it's possible to get these stars and M48 in the same field of view.

32

SPRING TARGETS

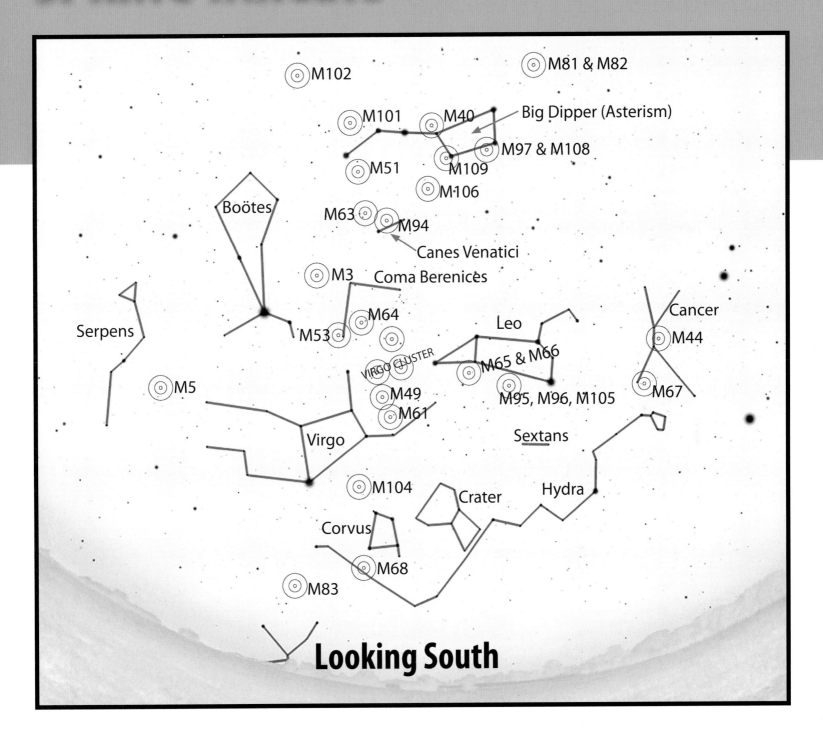

M102

M81 & M82

M101 M40 Big Dipper (Asterism)

M51 M97 & M108

M109

M106

Boötes

M63 M94

Canes Venatici

M3 Coma Berenices

Serpens

M64

Leo

Cancer

M53

M44

VIRGO CLUSTER

M65 & M66

M5

M49

M67

M61

M95, M96, M105

Virgo

Sextans

M104

Crater

Hydra

Corvus

M68

M83

Looking South

Stargazing in Spring

As the weather gets warmer and the snow begins to melt, amateur astronomers are more eager than ever to get back behind the eyepiece. Lengthening days, however, tend to push back John's stargazing sessions until after his kids are tucked into bed.

In contrast to our list of winter Messier objects, the springtime objects are almost all galaxies. This is primarily due to the 16 Messier galaxies in the Virgo Cluster, as well as 10 galaxies surrounding the Big Dipper. Observing and identifying the Virgo Cluster galaxies requires you to develop a new skill, "galaxy-hopping." Instead of navigating by the stars, you'll use reference galaxies to hop from one galaxy to the next.

The challenge with observing the springtime targets is that most of these galaxies require dark skies and moonless nights. Only a handful of these are visible from the suburbs, whether you're using a small refractor or a large Dobsonian.

From urban skies, the best spring targets are the open clusters M44 and M67. The Double Star M40 is an easy target, and globular cluster M3 can be spotted without too much difficulty.

The springtime sky is also home to some lesser known treasures. At star parties the planetary nebula M9, known as the Owl Nebula, isn't nearly as popular as M57, the Ring Nebula. But it should be, since it shares the field of view with galaxy M108, which becomes visible under dark sky conditions.

Depending on your latitude, M83 (the Southern Pinwheel Galaxy) and M68 (a globular cluster) may well be the most challenging targets this season. At around 45 degrees north latitude, these two objects hardly get more than an outstretched hand's length above the horizon.

Remember that many of the winter objects will still be visible in the early evening. If you're looking for more targets, try to find Melotte 111 (an open star cluster visible from the city). From darker skies, try to fit Melotte 111 and the Needle Galaxy (NGC 4565) into the same field of view.

Unable to leave the city due to the COVID-19 pandemic, John spent much of March 2021 working through the RASC's Explore the Moon program with two beginner telescopes.

M44

Dim stars make the constellation of Cancer somewhat difficult to identify, but once you've found it, the Beehive Cluster (M44) lies between its two central stars. From dark skies, you should have little difficulty seeing this cluster's brighter stars with the unaided eye. M44 borders the winter constellations, and come June, this cluster falls victim to the lengthening days.

M44's central shape reminds John of a beehive smoker his grandfather, Dr. Dean Read, used to use. (Dr. Read was a scientist who studied insects.)

Common Name(s): The Beehive, Praesepe
Type: Open Cluster
Brightness (Visual Magnitude): 3.1
Distance (Light-Years): 577
Difficulty (Subjective): 1

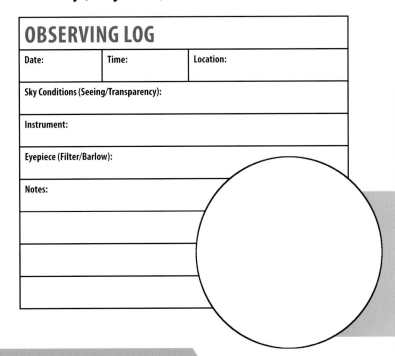

OBSERVING LOG

Date:	Time:	Location:
Sky Conditions (Seeing/Transparency):		
Instrument:		
Eyepiece (Filter/Barlow):		
Notes:		

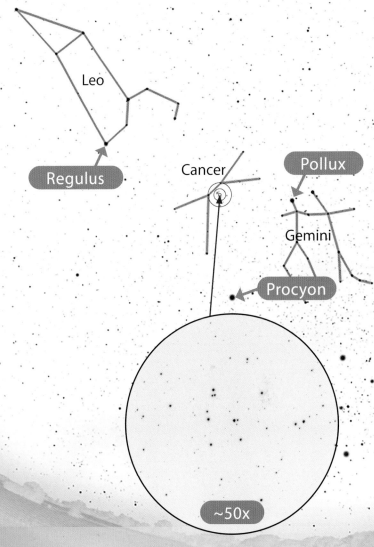

~50x

Observing Tips: M44 can be a great target for city viewing through a telescope or binoculars. But if you can't make out the stars of Cancer due to urban light pollution, find M44 by scanning the sky between the bright stars Regulus and Pollux. This target looks best at your lowest magnification.

M67

M67 fills a much larger volume of space than the Beehive (M44), but it looks smaller in the sky because it is over 5 times as distant. It appears only slightly smaller than the full Moon and makes a great target for binoculars in a dark sky. Take a closer look with a telescope (even from the city). Did you notice its Golden Eye—the yellow star?

Search the interior of the cluster for a squat triangle of stars—the cobra's head. A string of slightly brighter stars are arranged in a winding shape. Or do you see another pattern?

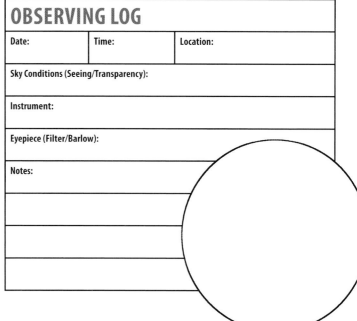

OBSERVING LOG

Date:	Time:	Location:

Sky Conditions (Seeing/Transparency):

Instrument:

Eyepiece (Filter/Barlow):

Notes:

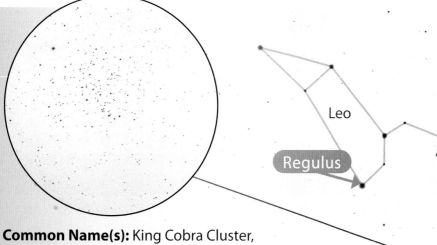

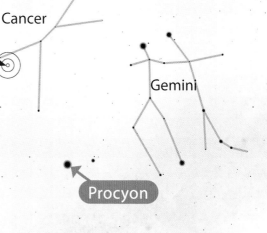

Common Name(s): King Cobra Cluster, Golden Eye Cluster
Type: Open Cluster
Brightness (Visual Magnitude): 6.9
Distance (Light-Years): 2,700
Difficulty (Subjective): 3

Observing Tips: Like the Beehive, this target is visible from urban locations in a telescope. Find the Beehive first and then pan the telescope slightly towards the horizon to find M67. Alternatively, search midway between Regulus and Procyon.

36

M40

You may find it odd that a double star is included in Messier's list of "objects that are not comets." Apparently, Charles Messier was searching for a previously reported nebula and found only these double stars in its place. And, as a diligent observer, Messier recorded his observation, even though he failed to identify something resembling a comet.

The telescope view image below shows a dim galaxy on the left-hand side of the frame. However, it's highly unlikely that Charles Messier could see this galaxy, named NGC 4290, with the telescope he was using at the time.

Common Name(s): Winnecke 4
Type: Double Star
Brightness (Visual Magnitude): 8.0
Distance (Light-Years): 510
Difficulty (Subjective): 2

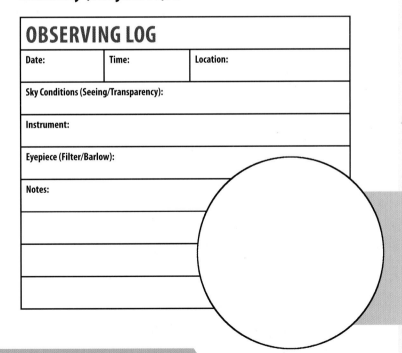

OBSERVING LOG

Date:	Time:	Location:
Sky Conditions (Seeing/Transparency):		
Instrument:		
Eyepiece (Filter/Barlow):		
Notes:		

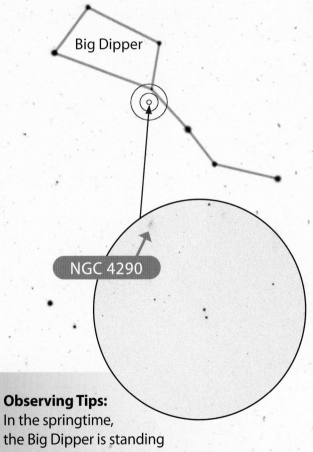

Big Dipper

NGC 4290

Observing Tips:
In the springtime, the Big Dipper is standing upright in the northeastern sky. To find M40, slightly extend the lower side of the bowl.

M81 & M82

The M81 and M82 galaxy pair are one of the most popular stargazing targets in the Northern Hemisphere. They are so bright that Chris can observe them from his driveway, under street lights! The larger spiral galaxy, M81, is usually spotted first, with the small, edge-on galaxy, M82, located in the same field of view.

M82 is classified as an irregular galaxy. It's also a "starburst" galaxy, referring to its above-average rate of star-formation driven by an abundance of available gas, and possibly by gravitational interactions with nearby M81.

These galaxies are circumpolar for much of the Northern Hemisphere. John views them during almost every session at the Burke-Gaffney Observatory, where he'll often set up five Dobsonian telescopes on the observation deck and then help the astronomy students find the most interesting deep-sky objects visible from the city.

Common Name(s): Bode's Galaxy (M81),
Cigar Galaxy (M82)
Type: Spiral Galaxy (M81), Irregular Galaxy (M82)
Brightness (Visual Magnitude): 6.9 (M81), 8.4 (M82)
Distance (Light-Years): 11.8 million (M81), 12.1 million (M82)
Difficulty (Subjective): 2

Observing Tips: Using two stars in the Big Dipper as a guide (as shown), M81 (and M82) are quite easy to find and can be viewed together in the same field of view at low magnification. In very dark skies, you may be able to see a third galaxy, NGC 3077.

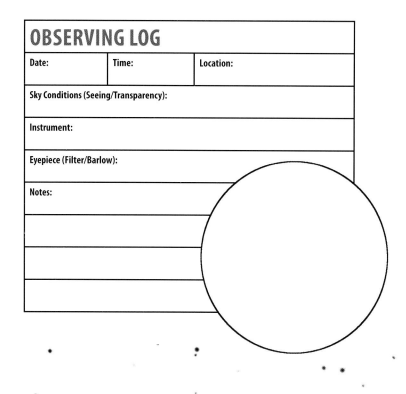

OBSERVING LOG

Date:	Time:	Location:

Sky Conditions (Seeing/Transparency):

Instrument:

Eyepiece (Filter/Barlow):

Notes:

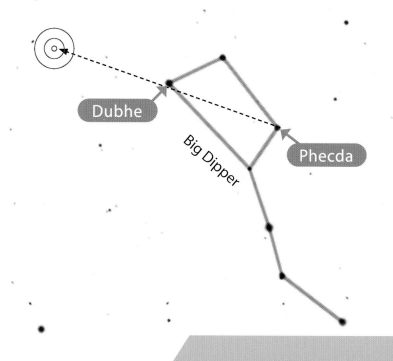

Dubhe

Big Dipper

Phecda

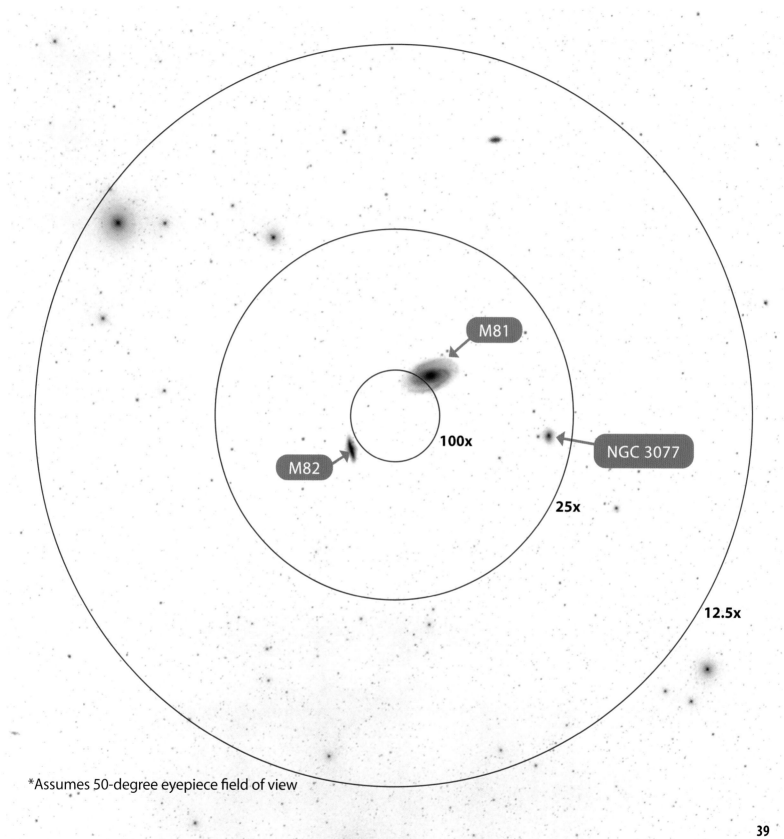

M81

M82

NGC 3077

100x

25x

12.5x

*Assumes 50-degree eyepiece field of view

M108 & M97

M108 requires dark skies. John tried on several occasions to see it from the city using a 24-inch PlaneWave CDK (a research-grade telescope) and was unsuccessful. From truly dark skies, finding this galaxy is relatively easy (assuming your eyes are dark adapted). If you're using a large enough telescope, you'll be sure to observe the finer structure within.

The planetary nebula, M97, is also best viewed under dark skies. If you look closely, you can make out the two large dark patches representing the eyes of an owl, which give this cluster its name.

Common Name(s): Surfboard Galaxy (M108), Owl Nebula (M97)
Type: Galaxy, Planetary Nebula
Brightness (Visual Magnitude): 10.0 (M108), 9.9 (M97)
Distance (Light-Years): 14.1 million (M108), 2,026 (M97)
Difficulty (Subjective): 5

Observing Tips: At low magnification, M108 shares the same field of view as the bright star Merak, a star in the bowl of the Big Dipper. Once centered on either target, the other will share the field of view. Use averted vision on these two. Once you find these, use higher power to view them individually. A narrowband filter will make the Owl (M97) pop.

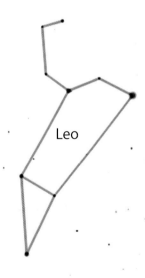

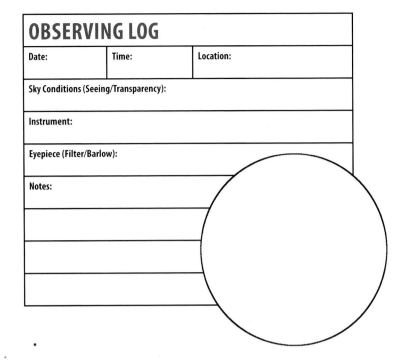

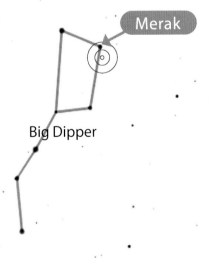

Merak

Big Dipper

Leo

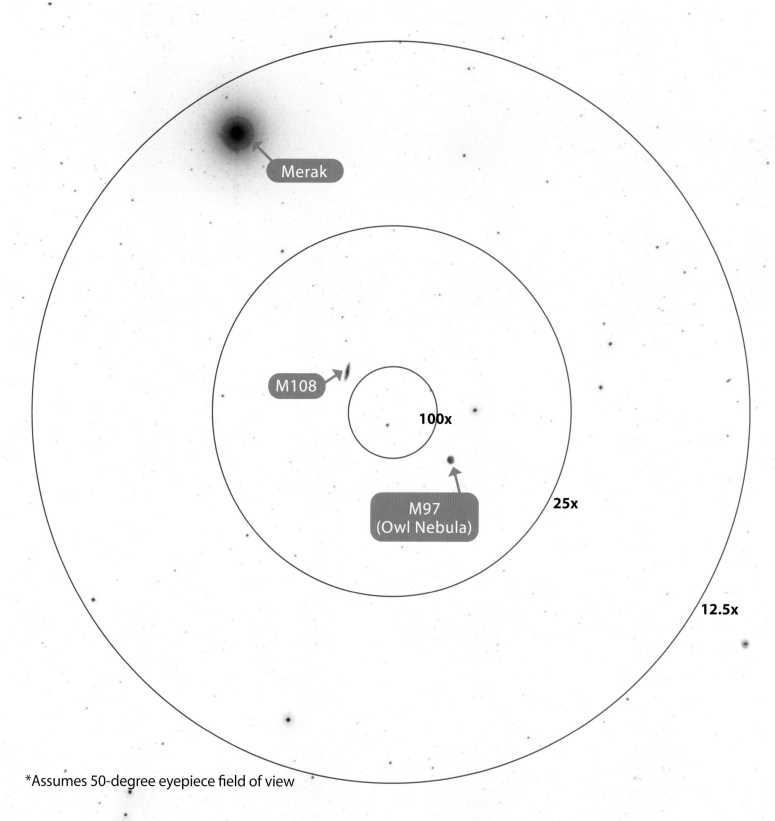

Merak

M108

100x

M97
(Owl Nebula)

25x

12.5x

*Assumes 50-degree eyepiece field of view

M109

M109 is one of the seven targets that Charles Messier made note of but never added to his published list. Through a telescope, even a large one, it's difficult for the eye to pick up more than this galaxy's central bar. This has to do with the extremely low surface brightness of the arms.

If your skies are dark and steady enough to clearly observe M109, check out its neighbor NGC 3953, a dimmer but similar looking galaxy located within the same field of view at around 30x power.

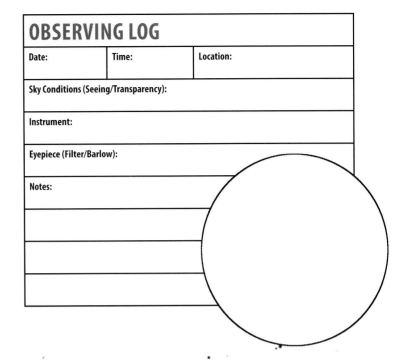

OBSERVING LOG

Date:	Time:	Location:

Sky Conditions (Seeing/Transparency):

Instrument:

Eyepiece (Filter/Barlow):

Notes:

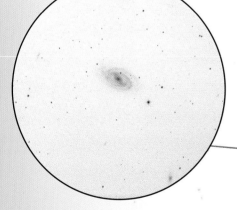

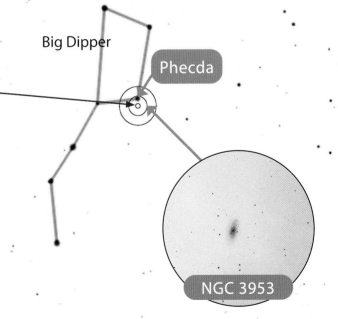

Common Name(s): Vacuum Cleaner Galaxy
Type: Barred Spiral Galaxy
Brightness (Visual Magnitude): 9.8
Distance (Light-Years): 83 million
Difficulty (Subjective): 4

Observing Tips: At low magnification, M109 will share the same field of view with the star Phecda. Use averted vision to get the best view of both M109 and NGC 3953.

Big Dipper

Phecda

NGC 3953

M106

M106 is brighter than M109 and may be possible to view from the suburbs depending on your sky conditions. This is a spiral galaxy, but it appears irregular through the eyepiece due to the contrasting features in the spiral arms. Under dark skies, use averted vision to see enhanced brightening on one side of the disk.

Pierre Méchain, Charles Messier's assistant and friend, is credited with the discovery of M106, as well as several other objects including eight comets. Méchain was 37 years old at the time of this discovery.

Common Name(s): None
Type: Spiral Galaxy
Brightness (Visual Magnitude): 8.4
Distance (Light-Years): 22.8 million
Difficulty (Subjective): 3

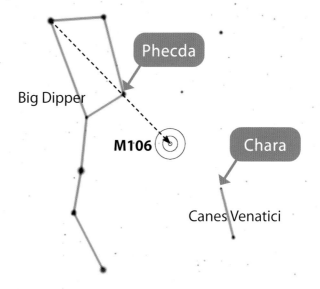

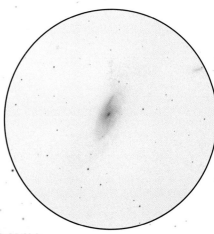

OBSERVING LOG

Date:	Time:	Location:

Sky Conditions (Seeing/Transparency):

Instrument:

Eyepiece (Filter/Barlow):

Notes:

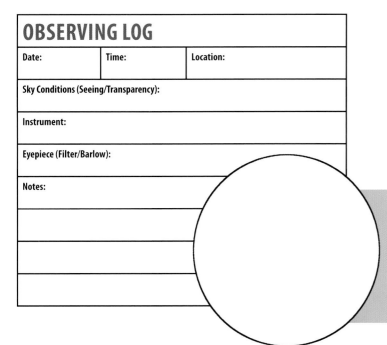

Observing Tips: Using the same stars you used to find M81 and M82, follow the same line in the opposite direction. M106 is located almost exactly halfway between Chara (a.k.a. Beta CVn) in Canes Venatici and Phecda in the Big Dipper.

M101

You'd think that at magnitude 7.9, M101 would be a prime target for the suburbs , especially considering that it's almost as large as the full Moon! But this is not the case! M101, like all galaxies, is an extended object and has a particularly low surface brightness. No matter how large your telescope, M101 will hide in even mildly light polluted skies.

For John, it wasn't until an outreach event in Yosemite National Park that this beauty came into focus in his brother-in-law's 8-inch Newtonian. If you have any amount of light pollution, it's better to focus on the nearby galaxy M51 instead, which is visible from the suburbs with less issue.

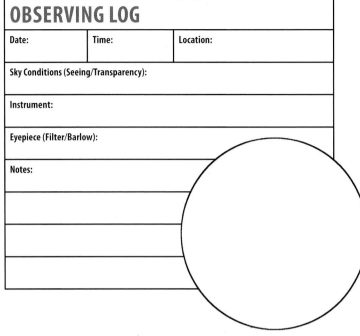

OBSERVING LOG

Date:	Time:	Location:

Sky Conditions (Seeing/Transparency):

Instrument:

Eyepiece (Filter/Barlow):

Notes:

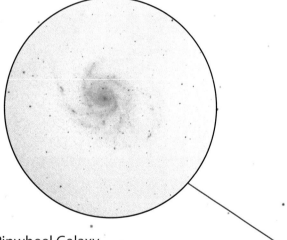

Common Name(s): Pinwheel Galaxy
Type: Spiral Galaxy
Brightness (Visual Magnitude): 7.9
Distance (Light-Years): 27 million
Difficulty (Subjective): 3

Observing Tips: From dark skies, this galaxy is easy to find, even with binoculars. Simply form an equilateral triangle with the two stars forming the end of the handle in the Big Dipper.

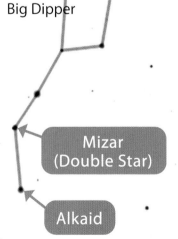

Big Dipper

Mizar (Double Star)

Alkaid

M51

These two conjoined galaxies form arguably the most beautiful and intricate structure visible in backyard telescopes. M51 appears as two smudges when viewed from the city under favorable conditions. But, when viewed under dark skies, and with larger instruments, the spiral structure comes into view, connecting the cores of these conjoined galaxies.

Charles Messier considered M51 a "double [nebula]" but gave it only the single designation. The New General Catalogue (NGC) assigned NGC 5195 to the smaller of the pair.

Common Name(s): Whirlpool Galaxy
Type: Spiral Galaxy, Irregular Galaxy (NGC 5195)
Brightness (Visual Magnitude): 8.4
Distance (Light-Years): 23 million
Difficulty (Subjective): 2

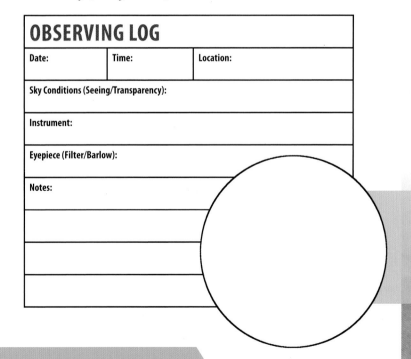

OBSERVING LOG

Date:	Time:	Location:
Sky Conditions (Seeing/Transparency):		
Instrument:		
Eyepiece (Filter/Barlow):		
Notes:		

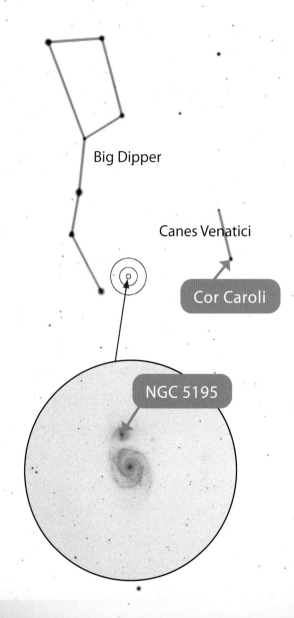

Big Dipper

Canes Venatici

Cor Caroli

NGC 5195

Observing Tips: M51 is a relatively easy target to find, even from the suburbs. From a dark site, this galaxy is visible in binoculars. Find it by forming a right triangle with the last two stars in the Big Dipper's handle.

M63

M63 is a spiral galaxy with a bright nucleus. At high magnification, its spiral arms are observed to be almost "chunky" in nature. A bright foreground star from our own galaxy sits towards one end of the disk.

When John took the telescope view image for the galaxy on April 19, 2017, he got an email the following morning from the director of the observatory. The image he'd taken contained a recently discovered and extremely dim supernova, designated SN 2017dfc. The supernova wasn't one of M64's stars, but a star in a background galaxy 40 times farther away.

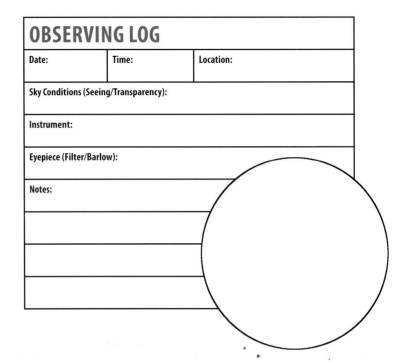

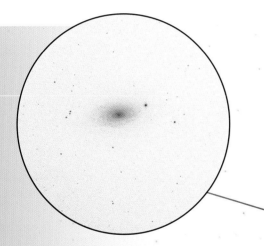

Common Name(s): Sunflower Galaxy
Type: Spiral Galaxy
Brightness (Visual Magnitude): 8.6
Distance (Light-Years): 37 million
Difficulty (Subjective): 3

Observing Tips: M63 is found on the line drawn between the bright stars Alkaid (tip of the Big Dipper's handle) and Cor Caroli, but somewhat closer to the latter. Under dark skies, use various eyepieces to play with the contrast in the spiral arms.

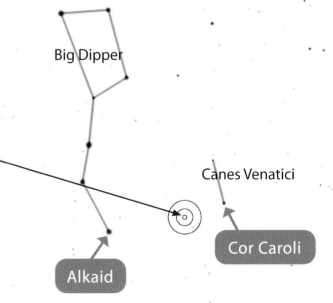

Big Dipper

Canes Venatici

Cor Caroli

Alkaid

M94

M94 is a spiral galaxy, but it looks like an elliptical galaxy or a planetary nebula in most amateur telescopes. It's not until you view it through large aperture optics that you can distinguish any structure.

While you're in the area, turn your telescope to Cor Caroli, a binary star system. The two component stars are bound together by gravity, but with an orbital period of several thousand years.

Common Name(s): Croc's Eye Galaxy
Type: Spiral Galaxy
Brightness (Visual Magnitude): 8.2
Distance (Light-Years): 16 million
Difficulty (Subjective): 3

OBSERVING LOG

Date:	Time:	Location:
Sky Conditions (Seeing/Transparency):		
Instrument:		
Eyepiece (Filter/Barlow):		
Notes:		

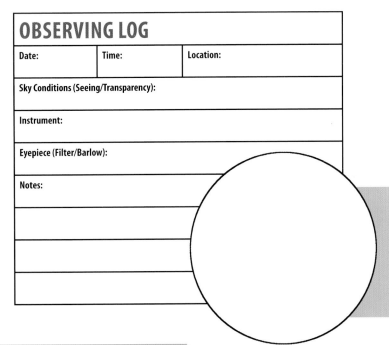

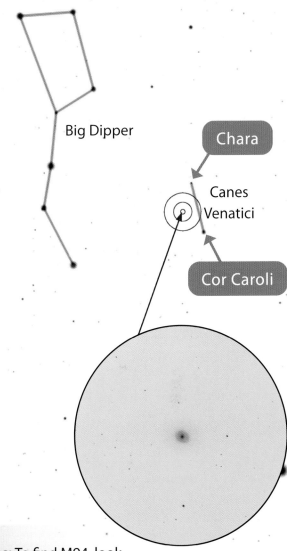

Big Dipper

Chara

Canes Venatici

Cor Caroli

Observing Tips: To find M94, look to the right of the handle of the Big Dipper and identify the two stars that make up Canes Venatici. This galaxy forms the top of a stout isosceles triangle with the two stars, Chara and Cor Caroli, as the base.

M65 & M66

M65 and M66, plus a third galaxy NGC 3628, form what is known as the Leo Triplet—one of two trios of galaxies in the constellation Leo.

In dark skies, the Leo Triplet is relatively easy to locate given its position just two finger widths below the Lion's hind leg. NGC 3628, nicknamed the Hamburger Galaxy for its interesting shape, was not included in Messier's list, perhaps due to its lower surface brightness. Sometimes when observing these galaxies under imperfect conditions, only M65 and M66 are visible.

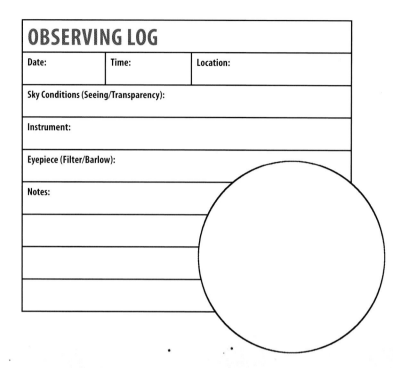

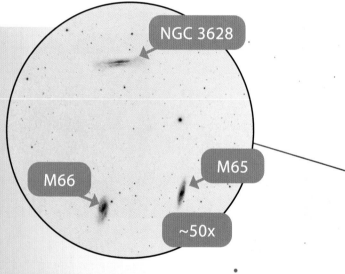

NGC 3628

M65

M66

~50x

Leo

Regulus

Common Name(s): Leo Triplet, M66 Group
Type: Spiral Galaxies
Brightness (Visual Magnitude): 9.3 (M65), 8.9 (M66)
Distance (Light-Years): 35 million
Difficulty (Subjective): 3

Observing Tips: We've found M65 and M66 challenging, but not impossible, to see from the city. On most nights they're simply overcome by light pollution. From dark skies, use low magnification to fit all three galaxies into the same field of view. Use a high-powered eyepiece to zoom in to better see their spiral structures.

M95, M96, M105

This tight trio of three Messier objects, M95, M96, M105 are the brightest members of a galaxy cluster called the M96 Group. If you're using a large telescope in dark skies, try to find other member galaxies in the same field of view.

We haven't had much luck seeing these galaxies from the city and the suburbs, only glimpsing M96 (barely) from the backyard with an 8-inch SCT telescope, but they're visible in any sized telescope from dark skies. At low magnification, all three galaxies will fit into the same field of view.

Common Name(s): Leo Trio, M96 Group
Type: Barred Spiral (M95), Spiral (M96), Elliptical (M105)
Visual Magnitude: 9.7 (M95), 9.2 (M96), 9.3 (M105)
Distance (Light-Years): ~32 million
Difficulty (Subjective): 3

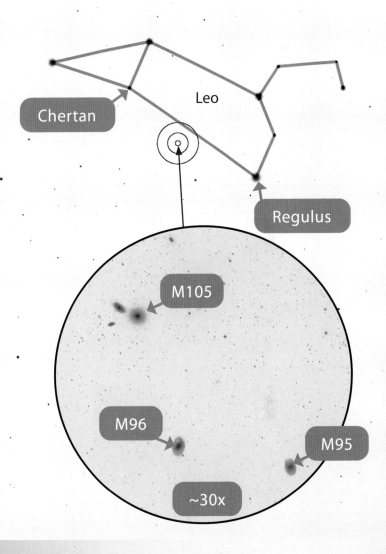

OBSERVING LOG

Date:	Time:	Location:

Sky Conditions (Seeing/Transparency):

Instrument:

Eyepiece (Filter/Barlow):

Notes:

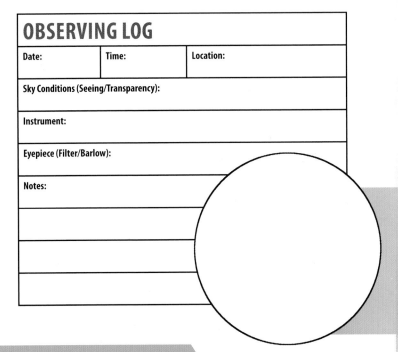

Observing Tips: To find the M96 group, form an imaginary line between Regulus and Chertan (a.k.a. Theta Leo). These galaxies can be found by setting the finder ring against the middle of this imaginary line, or by aiming the telescope about two finger widths below that imaginary line.

M53

Globular Cluster M53 sits within a finger's width of the star Diadem (a.k.a. Alpha Com) in the constellation Coma Berenices. From moderately dark skies, Coma Berenices is quite easy to find after "Arcing to Arcturus." But from the city, the three main stars in Coma Berenices disappear in the light pollution. It may take a bit of searching, but M53 can be observed through a telescope in light polluted skies.

Another much dimmer globular cluster, NGC 5053, can be spotted close by. Using low magnification, it's possible to view both clusters in the same field of view.

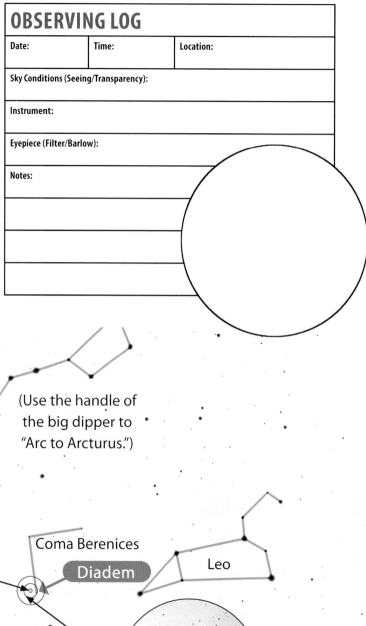

OBSERVING LOG

Date:	Time:	Location:

Sky Conditions (Seeing/Transparency):

Instrument:

Eyepiece (Filter/Barlow):

Notes:

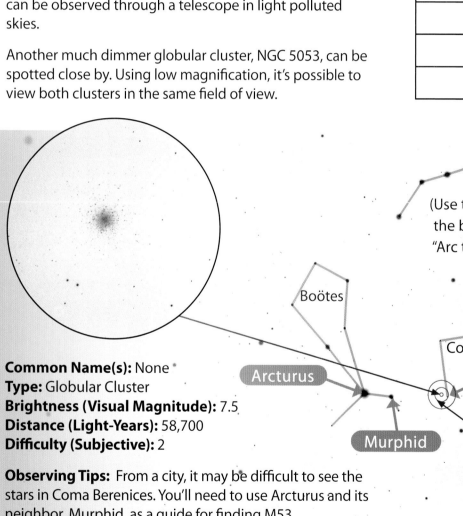

Common Name(s): None
Type: Globular Cluster
Brightness (Visual Magnitude): 7.5
Distance (Light-Years): 58,700
Difficulty (Subjective): 2

Observing Tips: From a city, it may be difficult to see the stars in Coma Berenices. You'll need to use Arcturus and its neighbor, Murphid, as a guide for finding M53.

(Use the handle of the big dipper to "Arc to Arcturus.")

Boötes

Coma Berenices

Arcturus

Diadem

Leo

Murphid

NGC 5053

M64

This galaxy is interesting because of a large dark dust band on one side of the galaxy's core. As you can see in the telescope view image, the "eye" in this galaxy's nickname is quite apparent, but it takes a larger telescope and dark skies to see this feature.

When viewing this part of the sky from the city, a better target would be Melotte 111, otherwise known as the Coma Star Cluster. This bright open cluster is visible even from light polluted skies with a small telescope or binoculars.

Common Name(s): Black Eye Galaxy, Sleeping Beauty Galaxy
Type: Spiral Galaxy
Brightness (Visual Magnitude): 8.5
Distance (Light-Years): 24 million
Difficulty (Subjective): 3

OBSERVING LOG

Date:	Time:	Location:
Sky Conditions (Seeing/Transparency):		
Instrument:		
Eyepiece (Filter/Barlow):		
Notes:		

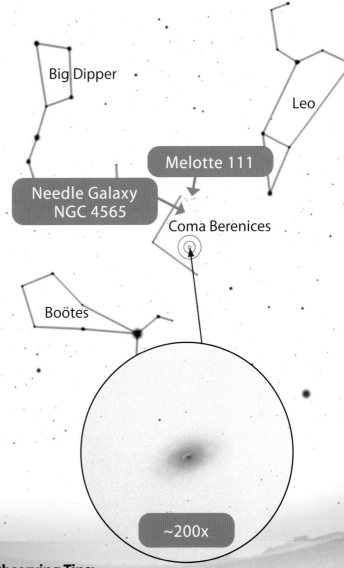

Big Dipper

Leo

Melotte 111

Needle Galaxy
NGC 4565

Coma Berenices

Boötes

~200x

Observing Tips:
M64 lies about one-third of the way along the hypotenuse of the triangle that makes up Coma Berenices. As with most galaxies, dark skies are required to see M64's details.

M3

M3 is one of the most beautiful globular clusters in our skies, appearing almost as brilliant as M13. What differentiates M3 from M13 is a triangle of three bright foreground stars surrounding it. From perfectly dark skies, this cluster can even be seen with the naked eye as a tiny patch of gray.

Telescopes with larger apertures easily show thousands of stars, even though Charles Messier described it as starless—a testament to just how much better telescopes have become.

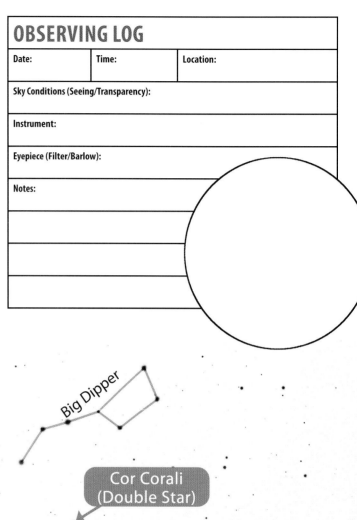

OBSERVING LOG

Date:	Time:	Location:

Sky Conditions (Seeing/Transparency):

Instrument:

Eyepiece (Filter/Barlow):

Notes:

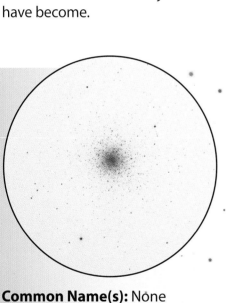

Common Name(s): None
Type: Globular Cluster
Brightness (Visual Magnitude): 5.9
Distance (Light-Years): 33,900
Difficulty (Subjective): 2

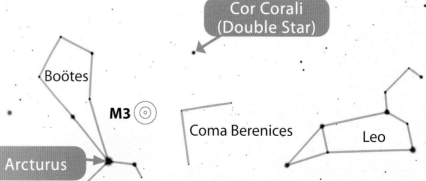

Observing Tips: M3 is bright enough to be observable even from a small city. It's located about midway between the very bright stars Arcturus and Cor Corali. John normally finds it by forming triangles with the bright stars in Boötes. In dark skies, it might be easier to extend the line formed by the two northerly stars in Coma Berenices.

Virgo Cluster Maps

The Virgo Cluster is a group of galaxies located within a plot of sky not much larger than your fist held at arm's length. This cluster of galaxies includes 16 Messier objects, all galaxies, shared between the constellations Virgo and Coma Berenices.

From dark skies, your telescope will reveal many more galaxies than Messier included. To correctly identify each Messier object, you'll need to hop from one galaxy to another, all while looking through your eyepiece.

For that reason, we've provided two sets of diagrams. The charts on the left page are for refractors and SCTs with diagonals. These charts show a mirror reversed view of the sky. The view on the right page shows a "standard view" to be used with image-erect telescopes, binoculars, and Newtonians.

Thirteen of the 16 Messier galaxies are split among two clumps—a group of 10 and a group of 3. Three additional Messiers are sprinkled around them. In the sky where the two bullseyes touch is a curved row of bright galaxies known as Markarian's Chain. This easy-to-see "landmark" provides an excellent anchor point for shifting your telescope slightly toward the northwest, where you'll find M98, M99, and M100.

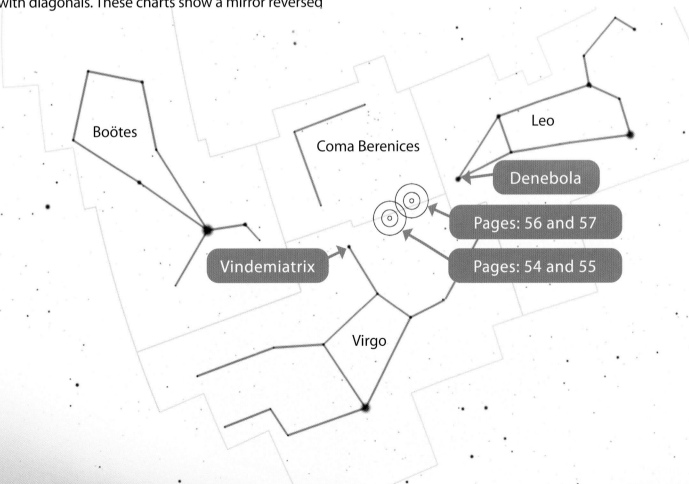

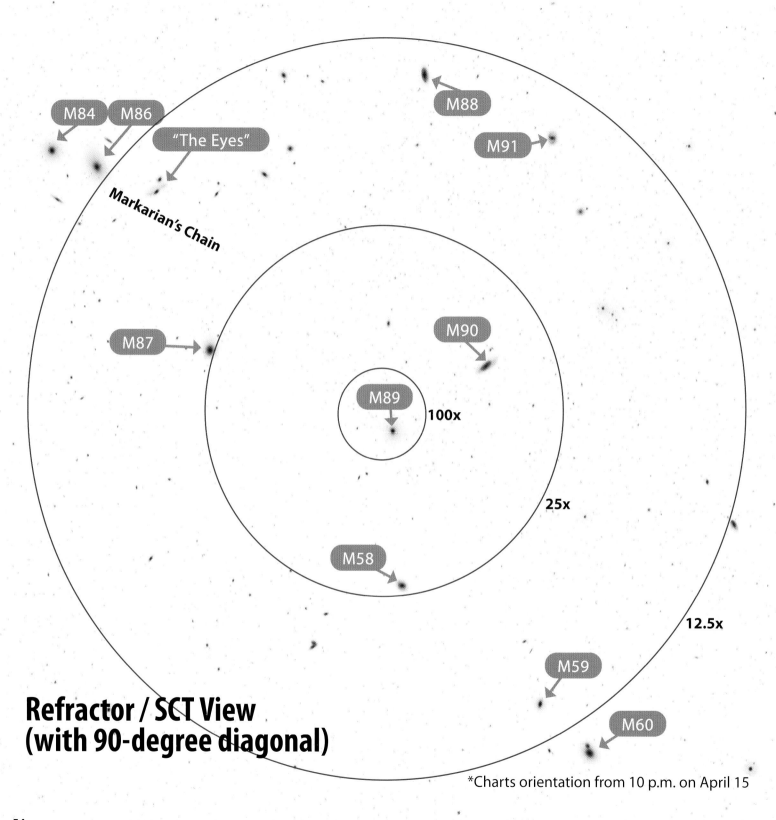

M84 M86

"The Eyes"

Markarian's Chain

M88

M91

M87

M90

M89

100x

M58

25x

12.5x

M59

M60

Refractor / SCT View (with 90-degree diagonal)

*Charts orientation from 10 p.m. on April 15

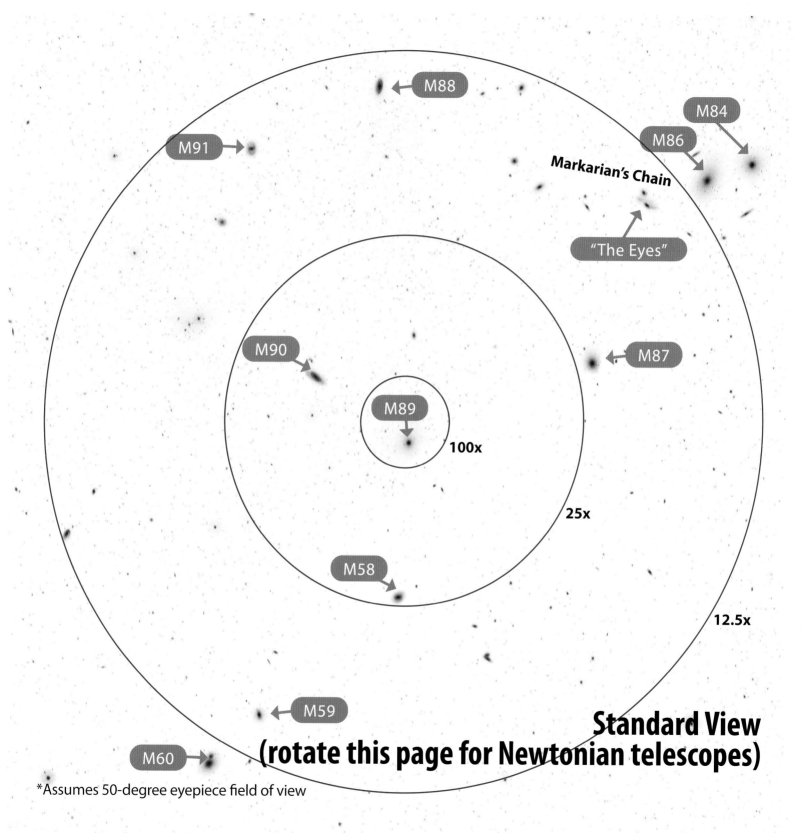

M88

M84

M86

M91

Markarian's Chain

"The Eyes"

M90

M87

M89

100x

25x

M58

12.5x

M59

**Standard View
(rotate this page for Newtonian telescopes)**

M60

*Assumes 50-degree eyepiece field of view

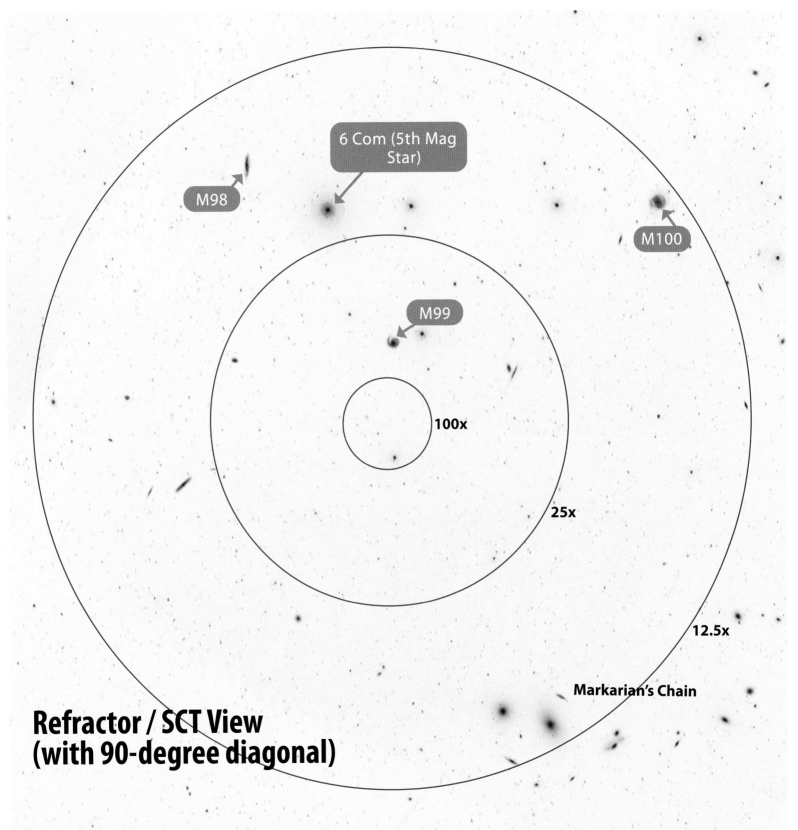

6 Com (5th Mag Star)

M98

M100

M99

100x

25x

12.5x

Markarian's Chain

**Refractor / SCT View
(with 90-degree diagonal)**

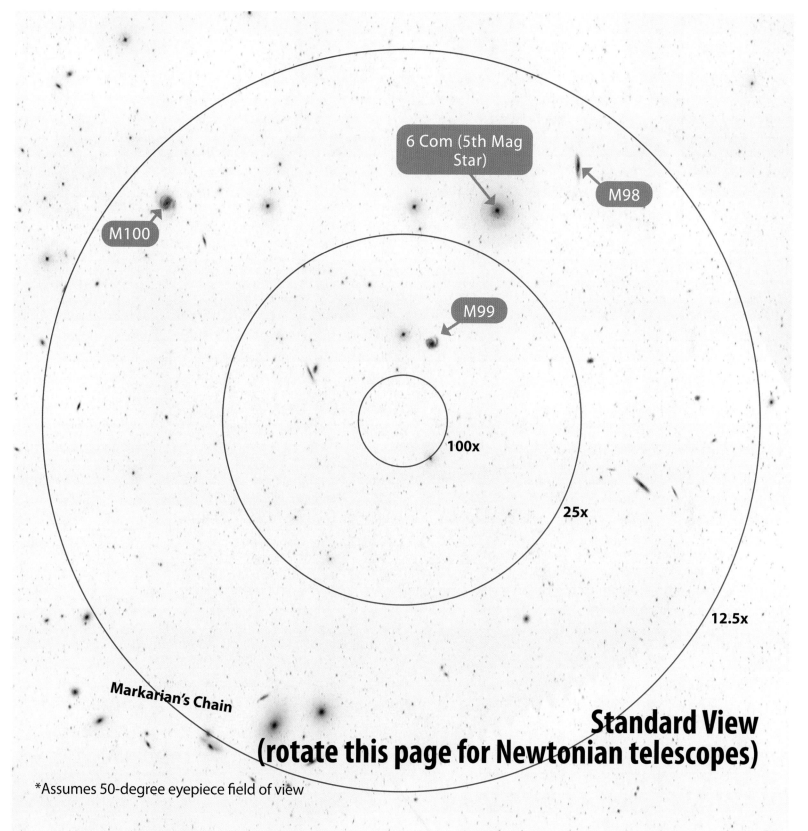

6 Com (5th Mag Star)

M98

M100

M99

100x

25x

12.5x

Markarian's Chain

**Standard View
(rotate this page for Newtonian telescopes)**

*Assumes 50-degree eyepiece field of view

Virgo Cluster Observation

Observing (and identifying) all the Messier objects in the Virgo Cluster is one of the most daunting tasks in completing the Messier list. But it's easier than you think—if you step through them in a logical order.

Observe these galaxies on a moonless night from a location far from city lights. A dark sky matters far more than the size of your telescope. In fact, many of these galaxies can be spotted with binoculars—by a seasoned observer under ideal conditions!

When planning your session, allow some time for your eyes to become fully dark adapted. A dim red flashlight will let you read these charts and write down your notes. Keep your phone or tablet tucked out of sight, or cover its screen with red film. To better see the fainter galaxies, use averted vision. Use a low-powered eyepiece and focus the telescope as well as you can.

Aim your telescope exactly midway between the bright stars Vindemiatrix in Virgo and Denebola. Your field of view should reveal two bright smudges, the big elliptical galaxies M84 and M86.

Those two galaxies sit at the western end of Markarian's Chain, a 1.5-degree long curved row of relatively bright galaxies! Each link of the chain is about the same width. To follow the chain, move from M84, the smaller galaxy, to M86, the brighter galaxy, and keep hopping.

The next galaxies are "the Eyes" (NGC 4435 and NGC 4438), a close-together pair of interacting galaxies oriented at right angles to the chain. A short distance farther along, you'll find another small, bright galaxy pair. From there the chain curves through two more links, each marked by a prominent elliptical galaxy.

Markarian's Chain is spectacular and highly recognizable, so it makes a terrific staging point for galaxy-hopping through the Virgo Cluster. If you get lost, just re-aim your telescope between Vindemiatrix and Denebola and start over.

Although most of the galaxies in the chain were originally discovered by Charles Messier and by William Herschel, Armenian astrophysicist Benjamin Markarian in the 1960s discovered that they are moving through the universe coherently.

Photo Marathon? Dark Sky NOT required!

Doing an Astrophotography Messier Marathon? Don't worry about those dark skies! Even the dimmest Messier objects tend to show up in a 30-second exposure if your telescope is properly tracking. Your photo may not make the cover of *Sky & Telescope* but at least you'll have fun in the process.

M84, M86, M87, M89

Date:	Time:	Location:
Sky Conditions:		
Instrument:		
Eyepiece:		

Use the galaxy-hopping maps on pages 54 and 55 while searching for these targets.

Name(s): M84, M86
Type: Elliptical Galaxies
Brightness: 9.1 (M84), 8.9 (M86)
Distance (LY): 60m (M84), 52m (M86)

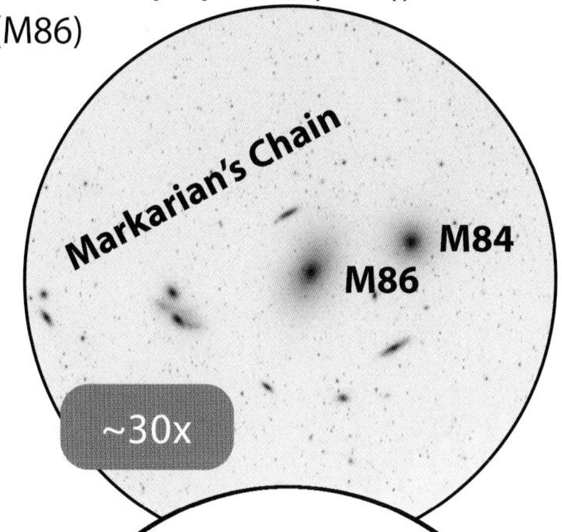

Name(s): M87
Type: Elliptical Galaxy
Brightness: 8.6
Distance (LY): 60 million

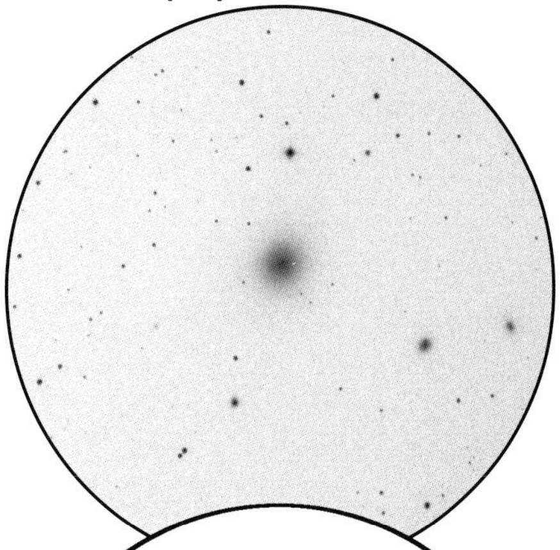

Name(s): M89
Type: Elliptical Galaxy
Brightness: 9.8
Distance (LY): 50 million

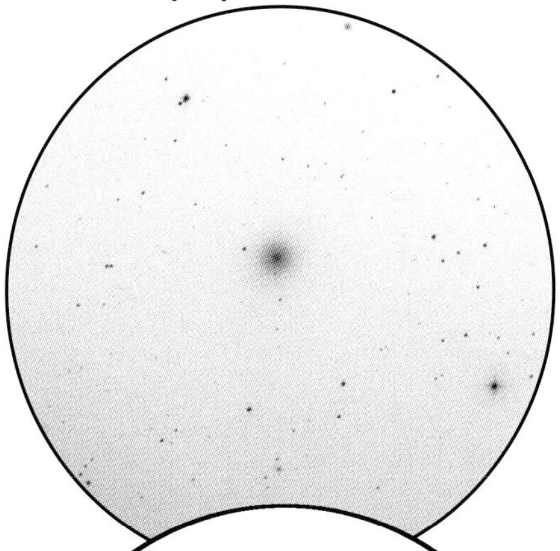

Notes:

M90, M91, M88

Use the galaxy-hopping diagrams on pages 54 and 55 while searching for these targets.

Date:	Time:	Location:
Sky Conditions:		
Instrument:		
Eyepiece:		

Name(s): M90
Type: Spiral Galaxy
Brightness: 9.5
Distance (LY): 59 million

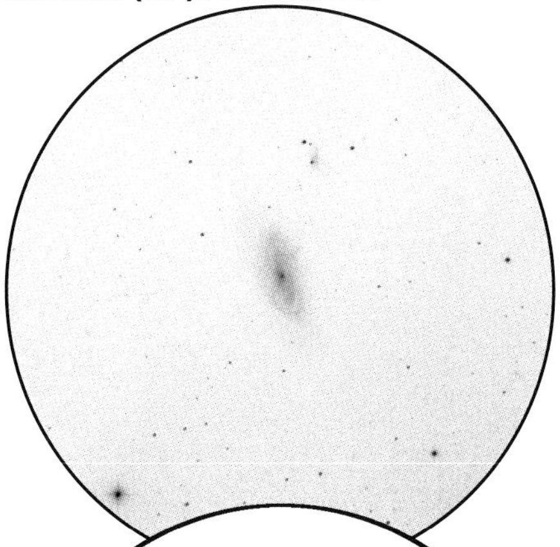

Name(s): M91
Type: Barred Spiral Galaxy
Brightness: 10.2
Distance (LY): 62 million

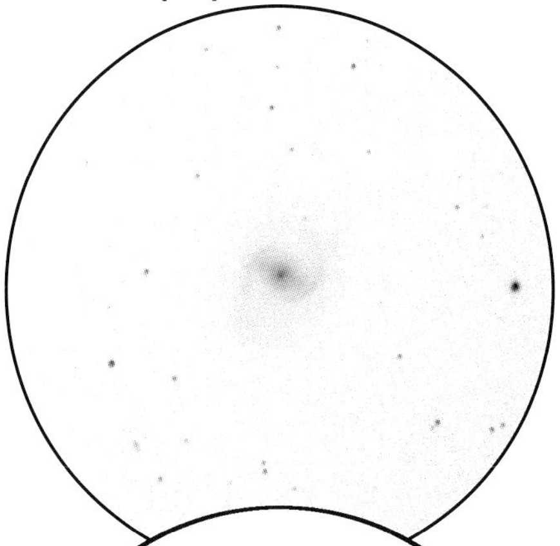

Name(s): M88
Type: Spiral Galaxy
Brightness: 9.5
Distance (LY): 47 million

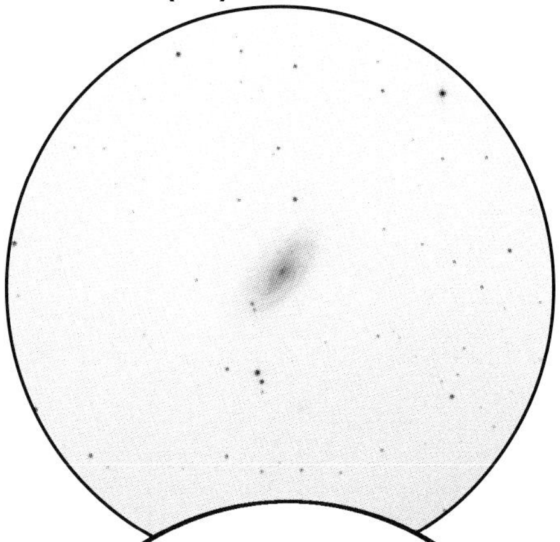

Notes:

M58, M59, M60

Use the galaxy-hopping diagrams on pages 54 and 55 while searching for these targets.

Name(s): M58
Type: Barred Spiral Galaxy
Brightness: 9.7
Distance (LY): 62 million

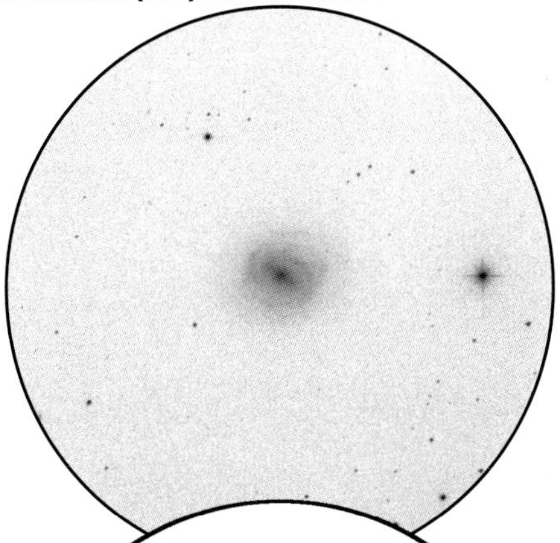

Name(s): M59
Type: Elliptical Galaxy
Brightness: 9.6
Distance (LY): 60 million

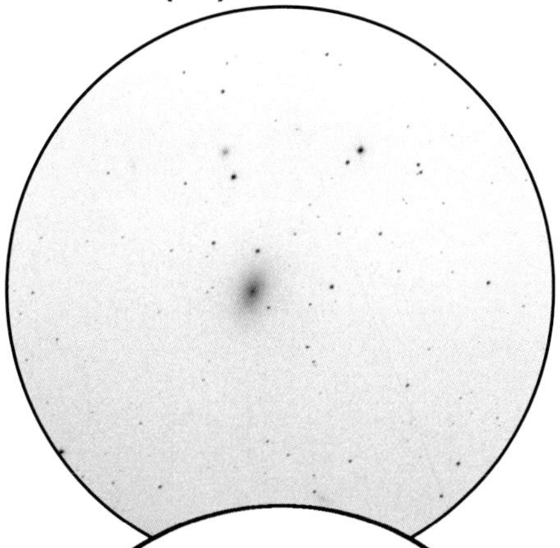

Name(s): M60
Type: Elliptical Galaxy
Brightness: 8.8
Distance (LY): 55 million

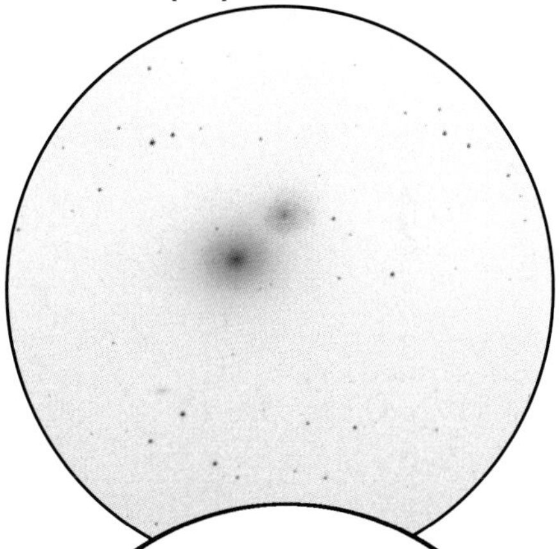

Notes:

M99, M98, M100

Date:	Time:	Location:
Sky Conditions:		
Instrument:		
Eyepiece:		

Use the galaxy-hopping diagram on pages 56 and 57 while searching for these targets.

Name(s): M99 (Virgo Pinwheel)
Type: Spiral Galaxy
Brightness: 9.9
Distance (LY): 60 million

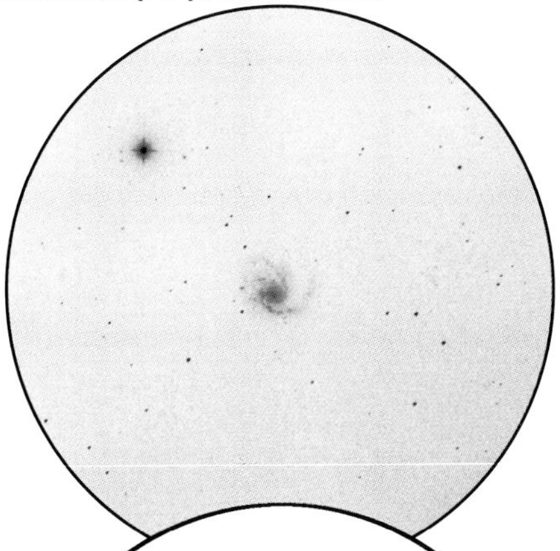

Name(s): M98
Type: Spiral Galaxy
Brightness: 10.1
Distance (LY): 57 million

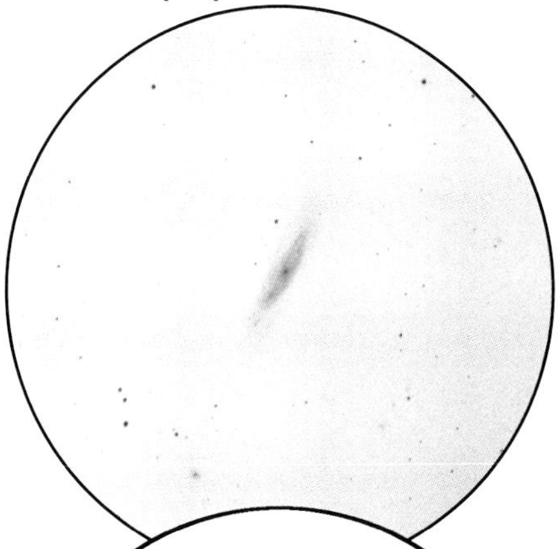

Name(s): M100 (Blowdryer Galaxy)
Type: Spiral Galaxy
Brightness: 9.4
Distance (LY): 55 million

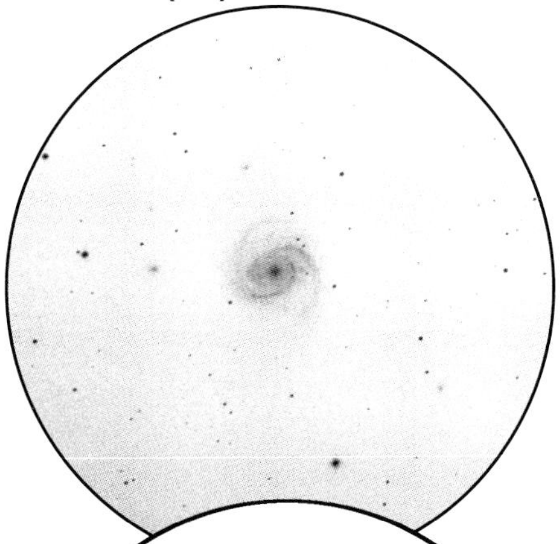

Notes:

Spring

M85

This galaxy lies at the apparent northern edge of the Virgo Cluster. M85 is huge, containing perhaps half a trillion stars. As with most elliptical or lenticular galaxies, M85 is largely devoid of star-forming gas and dust and appears as nothing more than an oval smudge—no matter what size of telescope you use.

Most targets get more interesting when you move up to larger and larger telescopes, but galaxies like M85 look pretty much the same no matter how powerful the instrument.

Common Name(s): M85
Type: Elliptical / Lenticular Galaxy
Brightness (Visual Magnitude): 9.1
Distance (Light-Years): 60 million
Difficulty (Subjective): 4

OBSERVING LOG

Date:	Time:	Location:

Sky Conditions (Seeing/Transparency):

Instrument:

Eyepiece (Filter/Barlow):

Notes:

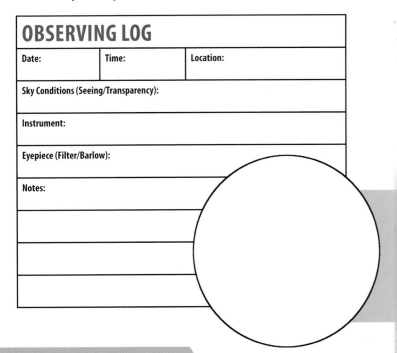

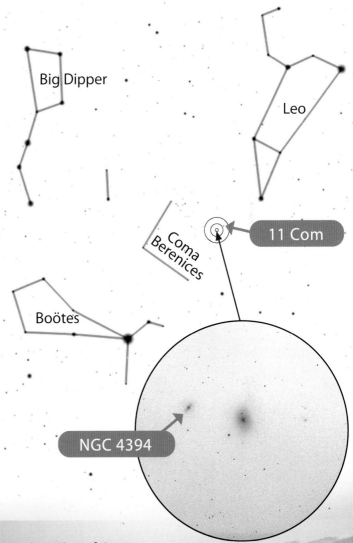

Observing Tips: If Coma Berenices were a box, M85 would be located in the corner near a dim (yet visible) star called 11 Com. The dim galaxy next to M85 in this eyepiece view is named NGC 4394.

M49

M49 is another large elliptical galaxy in the Virgo Cluster. Two smaller, nearby galaxies designated NGC 4535 and NGC 4526 can be seen in the same field of view at low magnification. They are nicknamed the "Lost Galaxies."

Through a telescope you'll notice a foreground star right next to the M49's nucleus. If this were any other galaxy, one might, at first glance, think that star was a supernova. Some supernovae hunters memorize the look of galaxies, revisiting them each night to look for these elusive exploding stars.

NGC 4526

NGC 4535

Arcturus

Leo

Virgo

M49

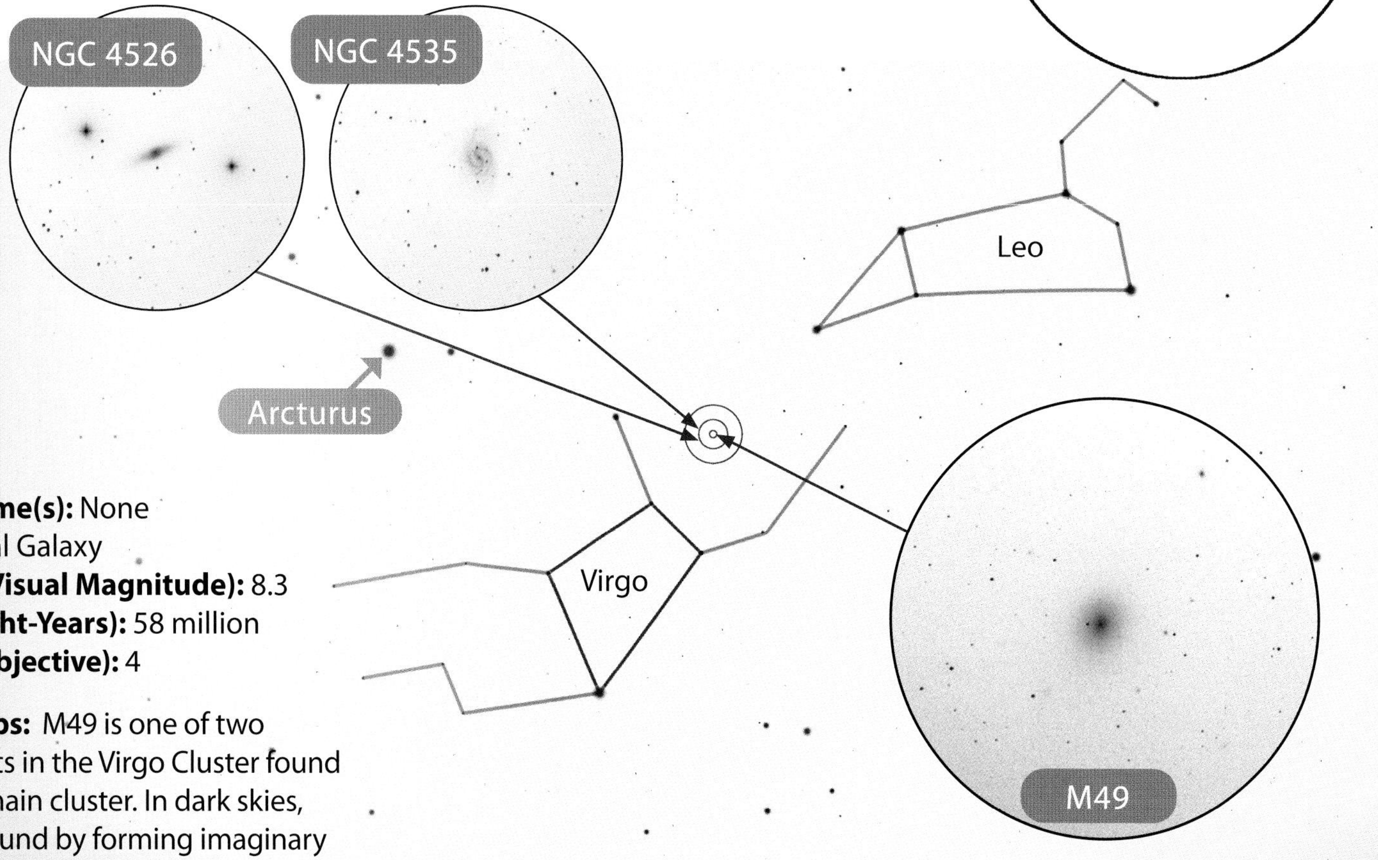

Common Name(s): None
Type: Elliptical Galaxy
Brightness (Visual Magnitude): 8.3
Distance (Light-Years): 58 million
Difficulty (Subjective): 4

Observing Tips: M49 is one of two Messier objects in the Virgo Cluster found south of the main cluster. In dark skies, M49 can be found by forming imaginary triangles with the stars in Virgo.

M61

We finish our tour of the Virgo Cluster's Messier objects with M61, a beautiful, face-on spiral galaxy. Under perfect seeing conditions and using a powerful telescope, a sea of dozens of dimmer background galaxies can be seen swimming above M61.

M61 was nicknamed the "Swelling Spiral" by author and astronomer Stephen O'Meara. The swelling refers to the apparent brightening and dimming of this galaxy's core as one switches between direct and averted vision.

Common Name(s): Swelling Spiral
Type: Spiral Galaxy
Brightness (Visual Magnitude): 9.7
Distance (Light-Years): 53 million
Difficulty (Subjective): 4

OBSERVING LOG

Date:	Time:	Location:
Sky Conditions (Seeing/Transparency):		
Instrument:		
Eyepiece (Filter/Barlow):		
Notes:		

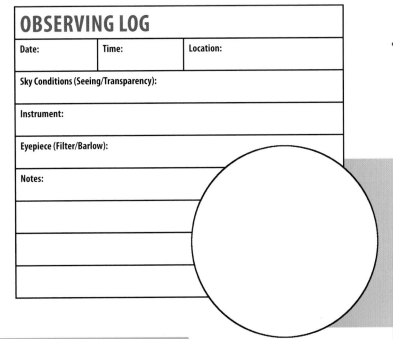

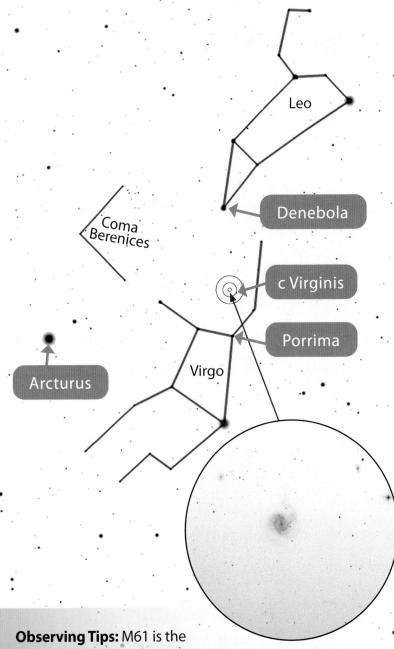

Observing Tips: M61 is the second of two Virgo Cluster Messier galaxies positioned well south of the main group. Find it along an imaginary line drawn between Denebola and Porrima, close to a medium-bright star named c Virginis. Do you agree that the bright foreground stars around this galaxy make it look like a mini solar system?

M104

M104, the aptly named Sombrero Galaxy, is one that we typically have no issue viewing from the city. Perhaps this is because its light is concentrated in a narrow streak, accented by a dark band of gas and dust that makes this galaxy stand out against the background sky.

The best view of this galaxy John ever had was with his little 90 mm Maksutov telescope at a star party in St. John, New Brunswick (Canada). The amazing contrast provided by the little scope at high magnification is why this tiny telescope is affectionately nicknamed the "Mighty Mak."

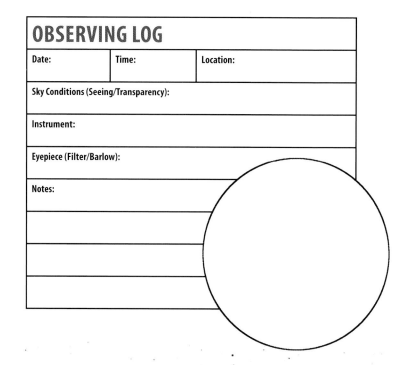

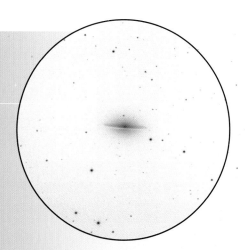

Common Name(s): Sombrero Galaxy
Type: Spiral Galaxy
Brightness (Visual Magnitude): 8.0
Distance (Light-Years): 29 million
Difficulty (Subjective): 3

Observing Tips: John usually finds M104 by forming an isosceles triangle with Spica and Porrima. At moderate magnification, a nearby asterism called "The Pistol" shares this galaxy's field of view. You'll know "The Pistol" when you see it!

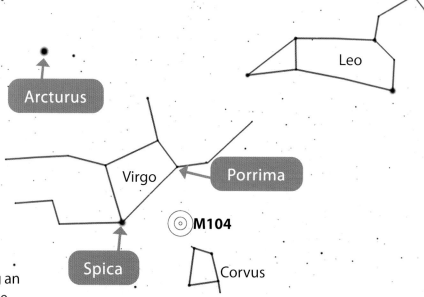

M68

While M68 is positioned within the constellation Hydra, the brighter stars in the diamond-shaped constellation Corvus make a much better reference point for finding this bright globular cluster, which can be viewed from light polluted skies.

For Northern Hemisphere observers, M68 sits quite low in the sky. From our locations, near 45 degrees north latitude, the cluster never rises more than 20 degrees above the horizon. The best time of night to search for M68 is when The Crow is "flying" highest over the southern horizon.

Common Name(s): None
Type: Globular Cluster
Brightness (Visual Magnitude): 7.7
Distance (Light-Years): 33,300
Difficulty (Subjective): 2

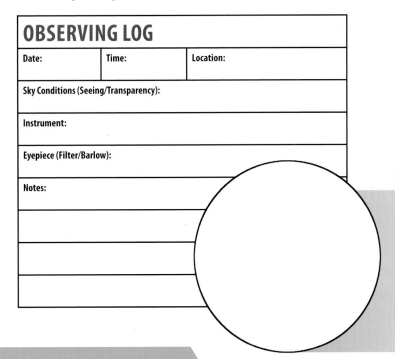

OBSERVING LOG

Date:	Time:	Location:

Sky Conditions (Seeing/Transparency):

Instrument:

Eyepiece (Filter/Barlow):

Notes:

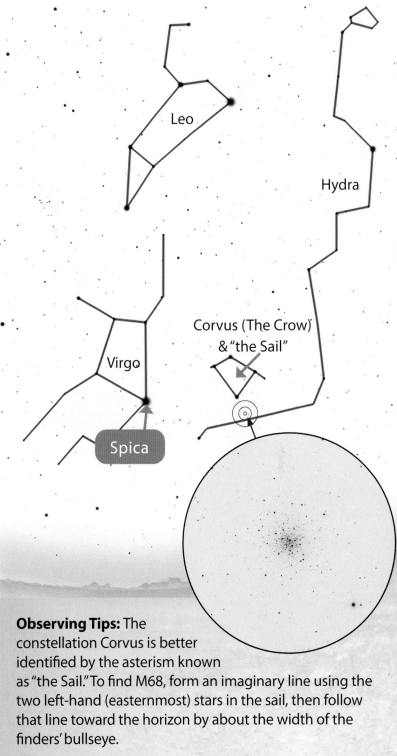

Observing Tips: The constellation Corvus is better identified by the asterism known as "the Sail." To find M68, form an imaginary line using the two left-hand (easternmost) stars in the sail, then follow that line toward the horizon by about the width of the finders' bullseye.

M102

There was some debate in the astronomy community as to whether M102 was no more than a duplicate reference to M101. However, evidence seems to support that NGC 5866 (pictured below) is the "real" M102. Messier's notes describe this object as located next to a "6th magnitude star." When searching for M102, the presence of a star of magnitude 5.2 (found west of Edasich) is indeed an effective way to identify this galaxy.

Views of this galaxy from dark skies in large instruments show a dark dust lane along the galaxy's plane, surrounded by a bright halo.

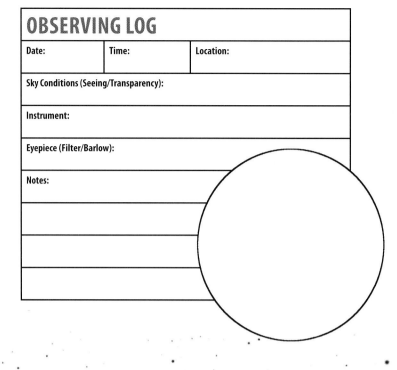

OBSERVING LOG

Date: Time: Location:

Sky Conditions (Seeing/Transparency):

Instrument:

Eyepiece (Filter/Barlow):

Notes:

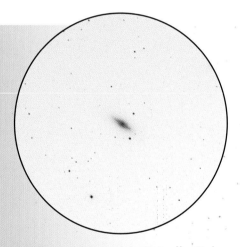

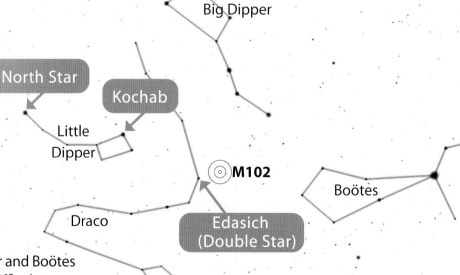

Common Name(s): Spindle Galaxy, NGC 5866
Type: Lenticular Galaxy
Brightness (Visual Magnitude): 9.9
Distance (Light-Years): 40 million
Difficulty (Subjective): 5

Observing Tips: We've added the Litter Dipper and Boötes to this map because the stars in Draco can be difficult to identify without additional reference points. M102 is located about a bullseye width west of the double star Edasich (a.k.a. Iota Dra).

M83

For John's location at about 45 degrees north, this face-on galaxy never rises much more than 15 degrees above the southern horizon. Even from sites farther south (he used to live in California), M83 can be a challenge to find due to its position in a featureless part of the elongated constellation of Hydra.

M83 is well worth searching for! This is one of the finest galaxies to see with a small telescope. In dark skies, its spiral arms are clearly visible.

Common Name(s): Southern Pinwheel
Type: Spiral Galaxy
Brightness (Visual Magnitude): 7.5
Distance (Light-Years): 15 million
Difficulty (Subjective): 4

OBSERVING LOG

Date:	Time:	Location:

Sky Conditions (Seeing/Transparency):

Instrument:

Eyepiece (Filter/Barlow):

Notes:

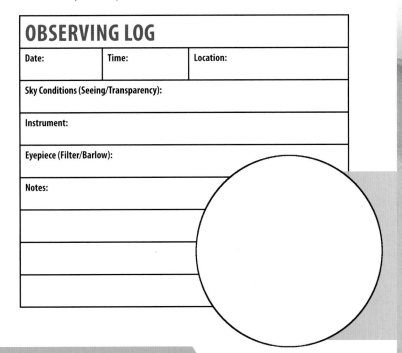

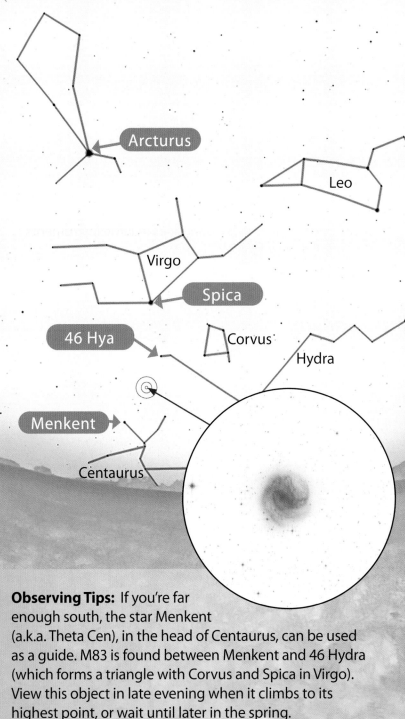

Observing Tips: If you're far enough south, the star Menkent (a.k.a. Theta Cen), in the head of Centaurus, can be used as a guide. M83 is found between Menkent and 46 Hydra (which forms a triangle with Corvus and Spica in Virgo). View this object in late evening when it climbs to its highest point, or wait until later in the spring.

M5

M5 is one of the brightest globular clusters visible in the Northern Hemisphere and is readily seen even from the city. John often uses it when comparing the quality of beginner telescopes. From perfectly dark skies, M5 can be seen without a telescope as a tiny smudge in the sky.

Its nickname the "Rose" comes from a halo of similarly bright stars arranged by chance in the shape of a rose. It reminds Chris of a face-on spiral galaxy! This effect doesn't show well in images, but rather as a trick of how the human eye perceives this cluster through the eyepiece.

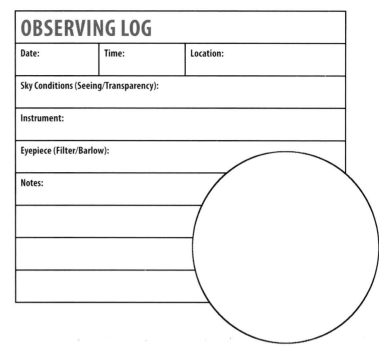

OBSERVING LOG

Date:	Time:	Location:

Sky Conditions (Seeing/Transparency):

Instrument:

Eyepiece (Filter/Barlow):

Notes:

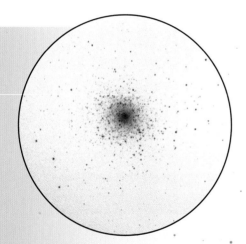

Common Name(s): Rose Cluster
Type: Globular Cluster
Brightness (Visual Magnitude): 5.7
Distance (Light-Years): 24,500
Difficulty (Subjective): 2

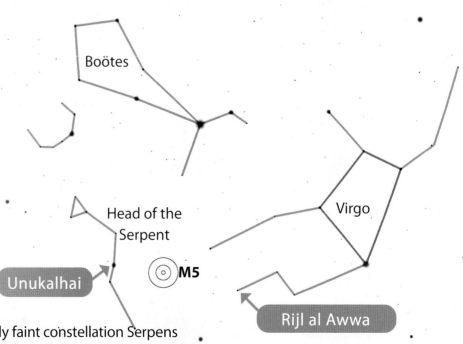

Observing Tips: M5 is located in the relatively faint constellation Serpens (the Snake), a constellation that is split into a western half near Boötes and Virgo and an eastern half beside Aquila, a summer/ fall constellation. M5 is in the "Head of the Serpent," along an imaginary line between the bright double star Unukalhai (a.k.a. Alpha Ser) and Riji al Awwa (a.k.a. Mu Vir) in Virgo.

SUMMER TARGETS

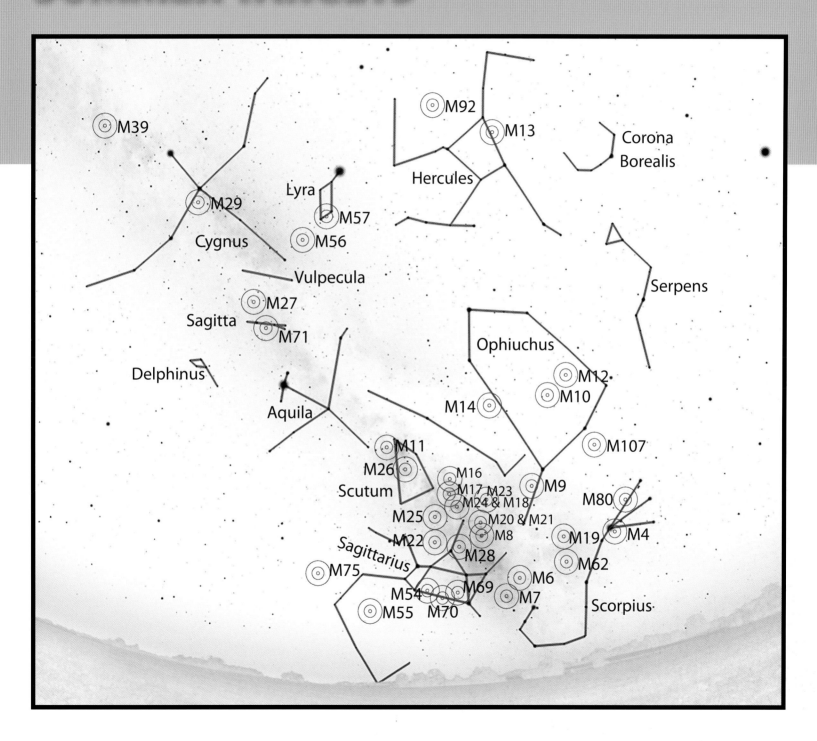

Stargazing in Summer

In addition to camping, beach bonfires, and road trips, stargazing is a quintessential summer experience. The challenges for stargazers during the summer are the short nights near the June solstice. The Sun simply doesn't get very far below the horizon, and the farther from the tropics you find yourself on Earth, the shorter the nights! This is why many summer star parties are held in late August. The organizers are simply waiting until the nights get a bit longer.

You may hear astronomers talk about astronomical twilight, which lasts from sunset until the Sun is 18 degrees below the horizon. Only after astronomical dusk (when twilight ends) is it considered "night" and the skies are truly dark.

At around 45 degrees north on June 21, dusk ends at around 11:30 p.m. and (astronomical) dawn starts at around 3:00 a.m.! This means that the night is only three and a half hours long! At more northerly latitudes, like in the UK for example, the Sun is NEVER more than 18 degrees below the horizon from mid-May to late July!

Summer stargazing is primarily focused around two separate parts of the sky: the Summer Triangle (an asterism connecting the stars Vega, Deneb, and Altair, and a second region near Sagittarius.

The most popular summer stargazing targets in the Messier list include M13 (the Great Globular Cluster in Hercules), M57 (the Ring Nebula), and M27 (the Dumbbell Nebula). Other popular targets found just above the southern horizon include M20 (the Trifid Nebula), M11 (the Wild Duck Cluster), and M8 (the Lagoon Nebula).

Other popular non-Messier summer stargazing targets include the Coathanger (a popular binocular target) and the Double Double (a.k.a. Epsilon Lyr, a double star in Lyra). Don't forget about circumpolar objects such as NGC 457 (the Dragonfly Cluster), which are visible all year long.

Summer also contains many popular non-Messier targets for astrophotograhers, including the Veil Nebula and the California Nebula.

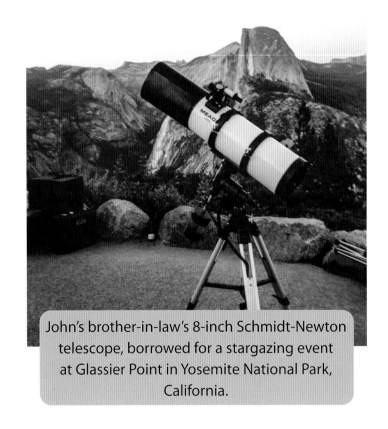

John's brother-in-law's 8-inch Schmidt-Newton telescope, borrowed for a stargazing event at Glassier Point in Yosemite National Park, California.

M4

M4 is one of the closest globular clusters to our Sun, and a great target for observing from the suburbs. This cluster's location beside Antares makes it the easiest Messier globular cluster to find.

The most memorable experience John had observing M4 was in a friend's backyard in Concord, California. After setting up the telescope on a concrete "island" in the middle of his backyard pool, his friend's three kids lined up to view this object. It was the first time they'd ever used a telescope. The kids had a great experience, and only one of them fell in the pool in the process.

Common Name(s): Crab Cluster
Type: Globular Cluster
Brightness (Visual Magnitude): 5.9
Distance (Light-Years): 7,200
Difficulty (Subjective): 2

Antares

Scorpius

OBSERVING LOG

Date:	Time:	Location:

Sky Conditions (Seeing/Transparency):		

Instrument:

Eyepiece (Filter/Barlow):

Notes:

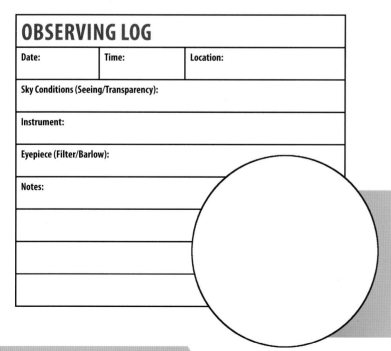

Observing Tips: At low magnification (around 25x), M4 will fit into in the same field of view as the bright red star Antares. However, bright Antares will make fainter M4 harder to see, so narrow the field of view by increasing the magnification once you've found it.

M80

M80 is a very small and dense globular cluster. It was actually a challenge to get the camera settings right for this cluster's eyepiece view (pictured below), as the stars tend to blend together as a blob.

Use high magnification to see more of the cluster's stars and its non-circular shape. More magnification will also help increase contrast under light polluted skies. M80 was easy to see in Chris's 80 mm refractor from a rural site a two-hour drive from Toronto, Canada.

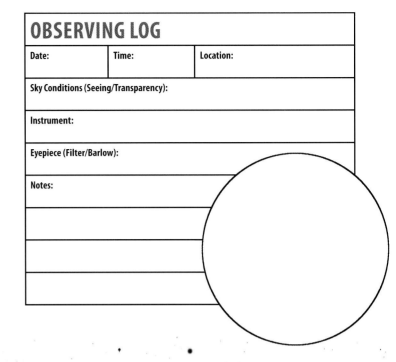

OBSERVING LOG

Date:	Time:	Location:

Sky Conditions (Seeing/Transparency):

Instrument:

Eyepiece (Filter/Barlow):

Notes:

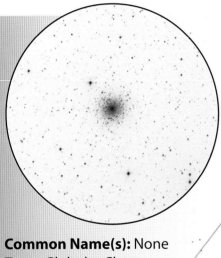

Common Name(s): None
Type: Globular Cluster
Brightness (Visual Magnitude): 7.3
Distance (Light-Years): 32,600
Difficulty (Subjective): 3

Observing Tips: M80 is found exactly halfway between the stars Acrab and Antares in Scorpius's claw. Acrab (a.k.a. Beta Sco), also known as Graffias, is a popular double star.

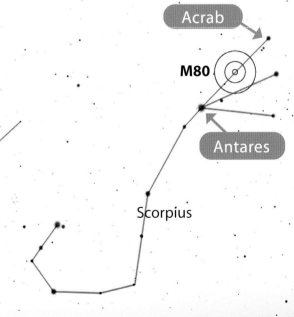

Acrab

M80

Antares

Sagittarius

Scorpius

From about 45 degrees north latitude, the base of Scorpius is on the horizon. This map shows a horizon from about 35 degrees north.

M62

M62 is found along the very southern boundary of the constellation Ophiuchus. This globular star cluster is located closer to the center of our Milky Way Galaxy than any other Messier object. Its brightness in visible wavelengths has been reduced by the gas and dust in the plane of our galaxy. This cluster was given its nickname by author Stephen O'Meara, who observed an optical illusion while observing the cluster at high magnification from a dark sky location. The cluster's core seemed to brighten and dim, almost flickering, as he shifted his gaze between the cluster's core and its halo.

Common Name(s): Flickering Globular Cluster
Type: Globular Cluster
Brightness (Visual Magnitude): 6.7
Distance (Light-Years): 22,200
Difficulty (Subjective): 3

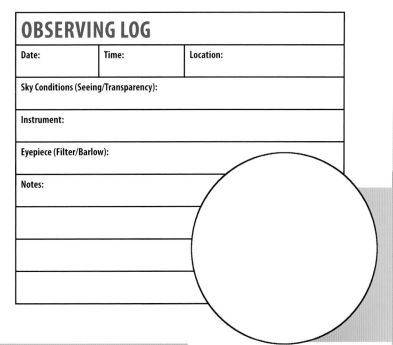

OBSERVING LOG

Date:	Time:	Location:

Sky Conditions (Seeing/Transparency):

Instrument:

Eyepiece (Filter/Barlow):

Notes:

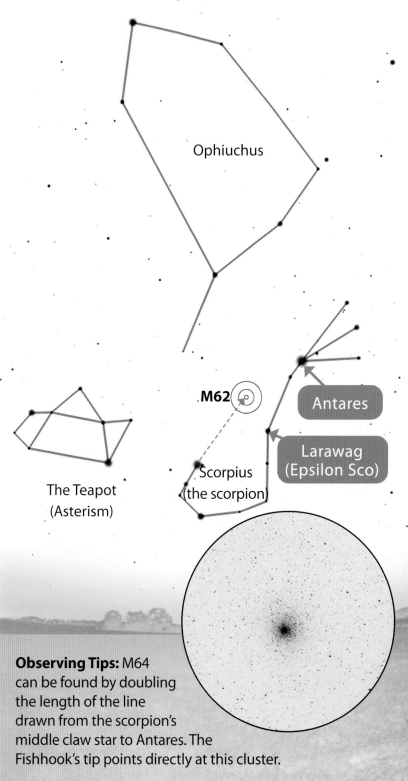

Ophiuchus

M62

Antares

Larawag
(Epsilon Sco)

Scorpius
(the scorpion)

The Teapot
(Asterism)

Observing Tips: M64 can be found by doubling the length of the line drawn from the scorpion's middle claw star to Antares. The Fishhook's tip points directly at this cluster.

M19

M19 can be quite a challenge to observe. It is one of the more distant globular clusters in Messier's list, and it is close to the galaxy's core, where foreground dust and gases reduce its luster.

M19 is located in Ophiuchus. Since the boundary of that constellation extends south toward the tip of Maui's Fishhook, an asterism in Scorpius, Ophiuchus actually straddles the ecliptic, making it a zodiac constellation. (Our zodiac signs should be Ophiuchus because the Sun was in this constellation when each of us was born.)

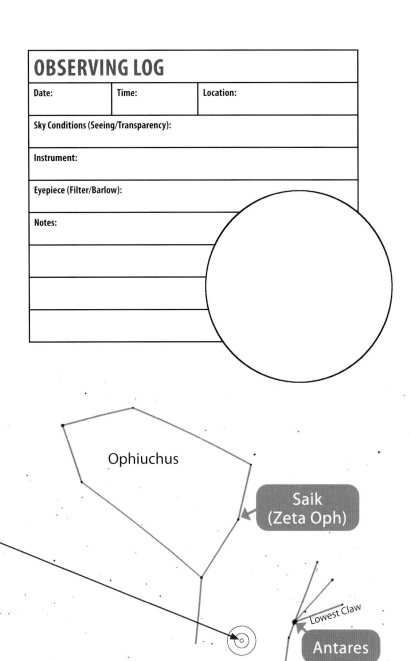

Common Name(s): None
Type: Globular Cluster
Brightness (Visual Magnitude): 6.7
Distance (Light-Years): 28,700
Difficulty (Subjective): 3

Observing Tips: To find M19, even from the suburbs, create a triangle to the upper left of the bright stars Antares and Larawag (a.k.a. Epsilon Sco) in Scorpius, or double the length of the line from the scorpion's lowest claw star to Antares.

M7

This scattering of blue stars, known as M7, is found within one of the most star-rich sections of the plane of our Milky Way Galaxy. Pictures don't do this cluster justice; it must be seen to be truly appreciated.

Binoculars and dark skies are the ideal way to view this cluster. Of course, it's always fun to dive in with a telescope as well. But use low magnification—this cluster is several times wider than the full Moon!

One July evening, Chris saw M7 with his unaided eyes from a Bortle 4 site north of Toronto, Canada.

Common Name(s): Ptolemy's Cluster
Type: Open Star Cluster
Brightness (Visual Magnitude): 3.3
Distance (Light-Years): 979
Difficulty (Subjective): 1

OBSERVING LOG

Date:	Time:	Location:
Sky Conditions (Seeing/Transparency):		
Instrument:		
Eyepiece (Filter/Barlow):		
Notes:		

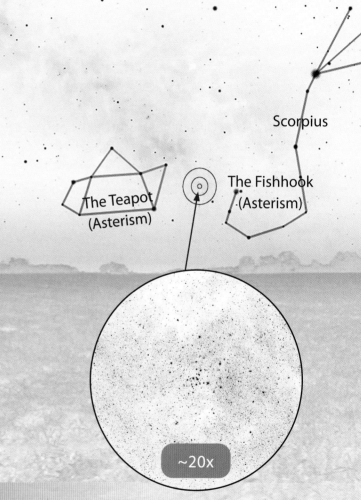

Scorpius

The Teapot
(Asterism)

The Fishhook
(Asterism)

~20x

Observing Tips: M7 is found a few degrees below M6, between the spout of the Teapot and the tip of the Fishhook. On a dark, moonless night, try to see this cluster with your naked eyes. This will be easier if you are located at a southern latitude.

M6

There are two open clusters that *really* look like what they are named after. One is NGC 457, the Dragonfly Cluster, in Cassiopeia. The other is M6, known as the Butterfly Cluster, which is just as delightful.

From dark skies, both M6 and nearby M7 are visible without a telescope, with M7 being the brighter of the pair. Every year (pandemics aside), there is an event called the Nova East Star Party, held at Smiley's Provincial Park in Nova Scotia. At this event, M6 and M7 are part of the "Ace Amateur Astronomer" program, binoculars targets for new stargazers created by ace astronomer Tony Schellinck.

~50x

The Teapot (Asterism)

M6

Scorpius

Shaula and Lesath

The Fishhook (Asterism)

Common Name(s): Butterfly Cluster
Type: Open Star Cluster
Brightness (Visual Magnitude): 4.2
Distance (Light-Years): 1,600
Difficulty (Subjective): 1

Observing Tips: M6 is found between the spout of the Teapot and the tip of the Fishhook (a visual double star, Shaula and Lesath). The only real challenge in viewing this cluster is finding a clear view of the southern horizon!

M9

M9 appears smaller and more condensed than the nearby clusters M10 (page 82) and M12 (page 81). That may be because it's farther away, or due to obscuring dust in the foreground (this cluster is located just above the galactic center). Of all the Messier objects in Ophiuchus, only M107 is dimmer, though it is arguable which cluster is more easily observed.

At very low magnification (around 25x), watch for two other globular clusters near M9. NGC 6356 is larger and brighter than M9, whereas NGC 6342 is quite a bit fainter.

Common Name(s): None
Type: Globular Cluster
Brightness (Visual Magnitude): 7.6
Distance (Light-Years): 25,800
Difficulty (Subjective): 2

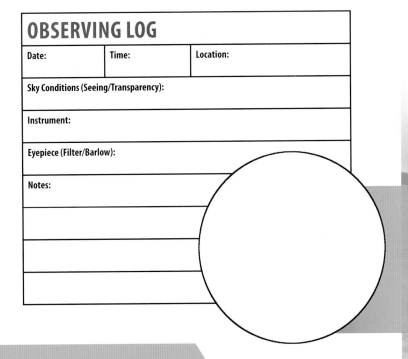

OBSERVING LOG

Date:	Time:	Location:

Sky Conditions (Seeing/Transparency):

Instrument:

Eyepiece (Filter/Barlow):

Notes:

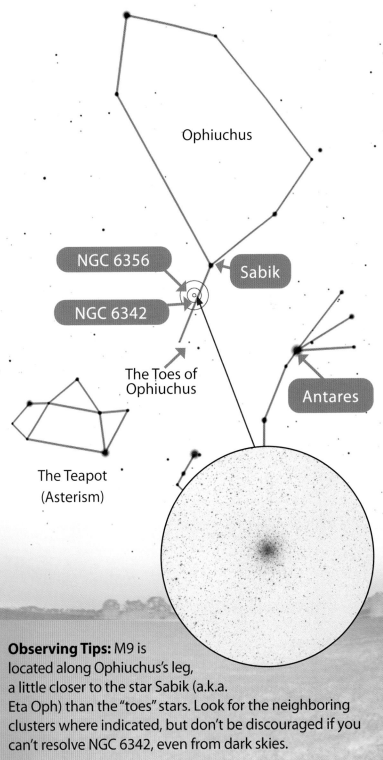

Ophiuchus

NGC 6356

NGC 6342

Sabik

The Toes of Ophiuchus

Antares

The Teapot (Asterism)

Observing Tips: M9 is located along Ophiuchus's leg, a little closer to the star Sabik (a.k.a. Eta Oph) than the "toes" stars. Look for the neighboring clusters where indicated, but don't be discouraged if you can't resolve NGC 6342, even from dark skies.

M107

M107 is another object not included in Messier's original list but added later (by Helen Sawyer Hogg) from Messier's notes. Hogg was the author of the classic stargazing book *The Stars Belong to Everyone*, a title that has become a popular catchphrase with the astronomy outreach community.

This globular cluster gets its nickname, the Crucifix Cluster, from the pattern made by the bright foreground stars that surround it. We've rotated the eyepiece view image below to show the cross-shape right-side-up.

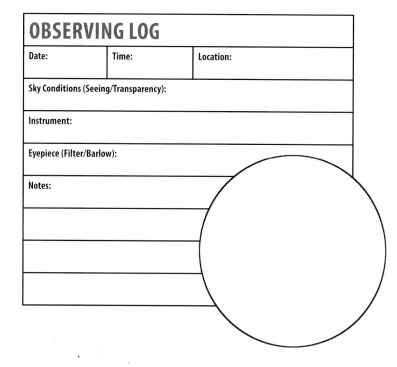

OBSERVING LOG

Date:	Time:	Location:

Sky Conditions (Seeing/Transparency):

Instrument:

Eyepiece (Filter/Barlow):

Notes:

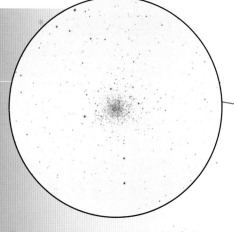

Common Name(s): The Crucifix Cluster
Type: Globular Cluster
Brightness (Visual Magnitude): 8.1
Distance (Light-Years): 20,900
Difficulty (Subjective): 2

Observing Tips: M107 is relatively easy to find by using Ophiuchus's star Saik (a.k.a. Zeta Oph), which isn't far from the very bright star Antares in Scorpius.

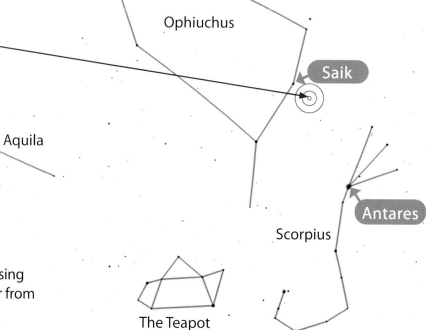

Ophiuchus

Saik

Aquila

Antares

Scorpius

The Teapot
(Asterism)

M12

Ophiuchus hosts several bright globular clusters, but M12 and M10 (on the next page) are the most brilliant. If you're doing a Messier Marathon in the spring, you'll hang out in this part of the sky while you wait for the Summer Triangle to climb a little higher in the sky.

Through a telescope at low magnification, author and astronomer Stephen O'Meara describes a rocket-shaped asterism located above M12, with the cluster representing the rocket's exhaust. What do you see?

Common Name(s): Gumball Globular Cluster
Type: Globular Cluster
Brightness (Visual Magnitude): 6.8
Distance (Light-Years): 15,600
Difficulty (Subjective): 2

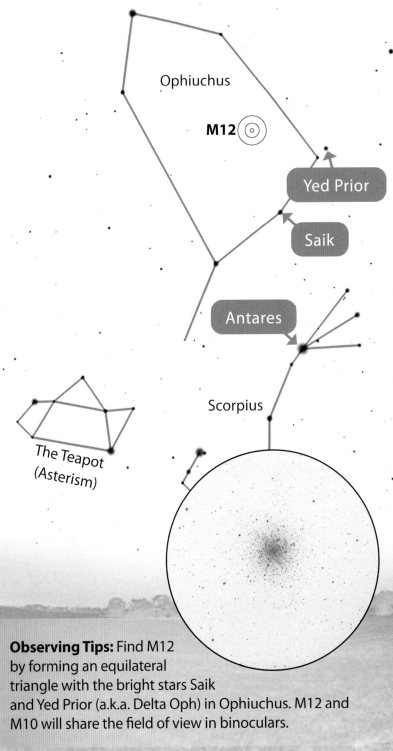

OBSERVING LOG

Date:	Time:	Location:
Sky Conditions (Seeing/Transparency):		
Instrument:		
Eyepiece (Filter/Barlow):		
Notes:		

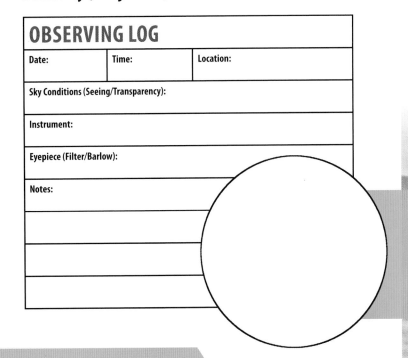

Observing Tips: Find M12 by forming an equilateral triangle with the bright stars Saik and Yed Prior (a.k.a. Delta Oph) in Ophiuchus. M12 and M10 will share the field of view in binoculars.

M10

The hexagon-shaped winter constellation Auriga features a row of three open star clusters, M36, M37, and M8. Coincidentally, Ophiuchus's (misshapen) hexagon contains three globular clusters in similar locations, M10, M12, and M14. Knowing that trick will let you quickly find three Messiers, no matter which season it is!

Only seven out of the many globular clusters hosted in Ophiuchus are on Messier's list. Most of the clusters are concentrated near the border with Scorpius, since that region of sky is located above the core of our Milky Way Galaxy.

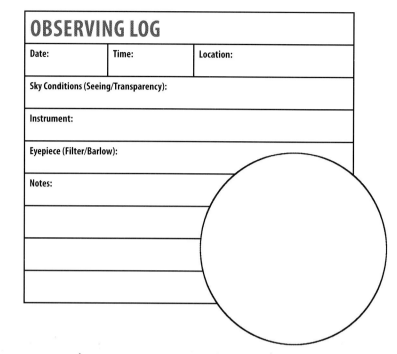

OBSERVING LOG

Date:	Time:	Location:

Sky Conditions (Seeing/Transparency):

Instrument:

Eyepiece (Filter/Barlow):

Notes:

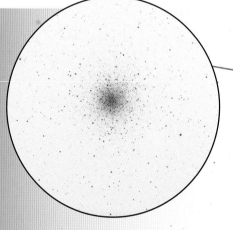

Common Name(s): None
Type: Globular Cluster
Brightness (Visual Magnitude): 6.6
Distance (Light-Years): 14,400
Difficulty (Subjective): 2

Observing Tips: M10 is located on the centerline of Ophiuchus's body. It can also be found by forming a right angle with Saik and Lambda Oph. At low magnification (around 25x), this cluster shares the field of view with a bright double-star system, 30 Oph.

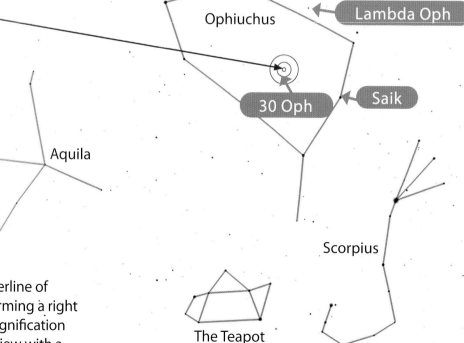

Ophiuchus

Lambda Oph

30 Oph

Saik

Aquila

Scorpius

The Teapot
(Asterism)

M14

M14 is a dense globular cluster situated just outside Ophiuchus's misshaped hexagon. It's rarely visited due to its obscure location, relatively far from bright stars and other Messier objects, but it's worth seeking out! Another globular cluster named NGC 6366 sits several finger widths to M14's lower right. Both clusters are easily visible in a small telescope from dark skies. NGC 6366 seems to be even more overlooked. Based on the database at the Burke-Gaffney Observatory, John was the first person to take a picture of this cluster with the telescope! For that reason, he's going to nickname this pair the "Lonely Cluster."

Common Name(s): Lonely Cluster
Type: Globular Cluster
Brightness (Visual Magnitude): 7.7
Distance (Light-Years): 29,000
Difficulty (Subjective): 2

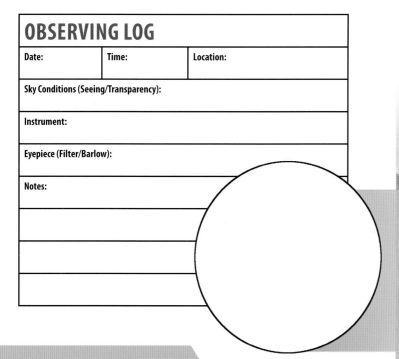

OBSERVING LOG

Date:	Time:	Location:

Sky Conditions (Seeing/Transparency):

Instrument:

Eyepiece (Filter/Barlow):

Notes:

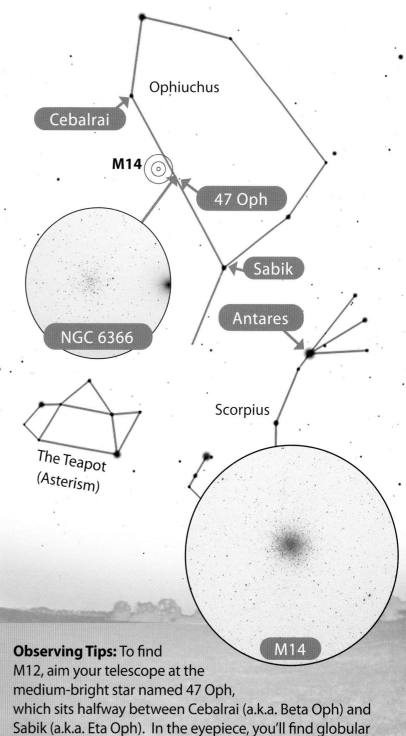

Observing Tips: To find M12, aim your telescope at the medium-bright star named 47 Oph, which sits halfway between Cebalrai (a.k.a. Beta Oph) and Sabik (a.k.a. Eta Oph). In the eyepiece, you'll find globular cluster NGC 6366 sitting beside 47 Oph. M14 is a short hop a few degrees to the northeast.

M20 & M21

The Trifid Nebula (M20) is one of the most photographed nebulae in the summer skies, and for good reason. M20 is a combination of different types of nebulae. A blue reflection nebula combines with a red emission nebula to produce purple in photos, and wisps of dark dust in the foreground divide the glowing gases into three sections. Use a narrowband filter to brighten M20.

While M20 is a very popular target, rarely do folks take the time to observe its neighbor, the open cluster M21. The name, Webb's Cross, doesn't technically refer to M21, but rather to an asterism of bright stars stretching between M20 and M21.

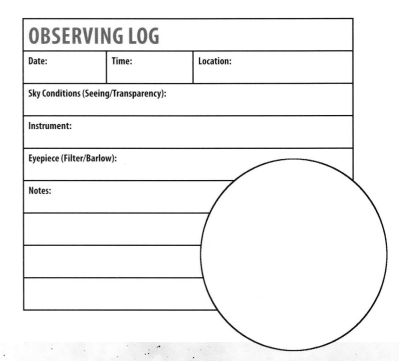

OBSERVING LOG

Date: Time: Location:

Sky Conditions (Seeing/Transparency):

Instrument:

Eyepiece (Filter/Barlow):

Notes:

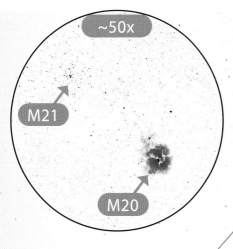

Common Name(s): Trifid Nebula (M20), Webb's Cross (M21)
Type: Emission Nebula (M20), Open Star Cluster (M21)
Brightness (Visual Magnitude): 6.3 (M20), 5.9 (M21)
Distance (Light-Years): 5,200 (M20), 4,200 (M21)
Difficulty (Subjective): 2

Observing Tips: Using around 50x magnification, both M20 and M21 fit nicely into the same field of view. The asterism Webb's Cross has M20 at the head and M21 at the foot. The arms are lopsided. These objects are found just west of Polis (a.k.a. Mu Sgr), the star at the top of Sagittarius.

M8

The Lagoon Nebula, M8, is a star-forming emission nebula covering an area of sky over three times as large as the full Moon. This cloud of bright gas is divided into two sections, separated by a band of dark nebulosity.

The bright region in the center (dark in the inverted image below) is known as the Hourglass. Chris was struck by how much larger and brighter the nebula became through his narrowband filter. The cluster of bright stars that appears within the nebula has a separate designation, NGC 6350.

Common Name(s): Lagoon Nebula
Type: Emission Nebula
Brightness (Visual Magnitude): 6.0
Distance (Light-Years): 5,100
Difficulty (Subjective): 2

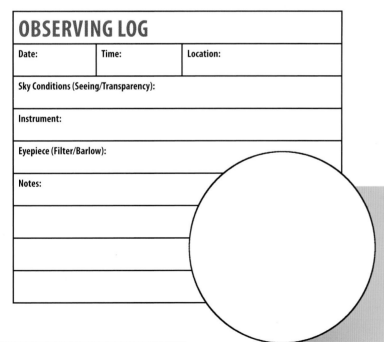

OBSERVING LOG

Date:	Time:	Location:

Sky Conditions (Seeing/Transparency):

Instrument:

Eyepiece (Filter/Barlow):

Notes:

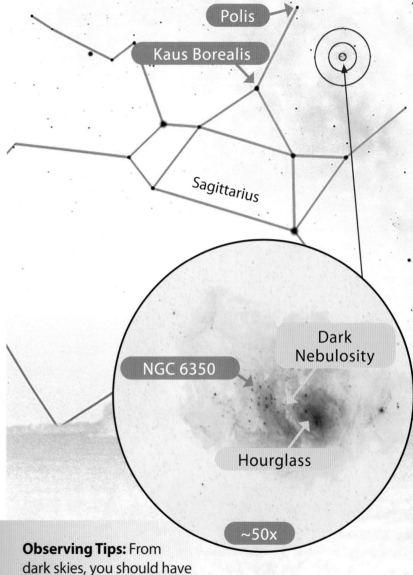

Polis

Kaus Borealis

Sagittarius

NGC 6350

Dark Nebulosity

Hourglass

~50x

Observing Tips: From dark skies, you should have no trouble seeing M8 without a telescope. It almost looks like someone is shining a flashlight on a distant cloud. It can also be found by following the steam (a bright patch of the Milky Way) from the spout of the teapot.

M24 & M18

The Small Sagittarius Star Cloud is a nearly invisible from the city but a masterpiece in dark skies. This bright clump of stars is located just above the star Polis, in the direction of Scutum's lower right corner. The Large Sagittarius Star Cloud was clearly not a comet to Messier's eyes. Many imagine it as the steam rising out of the Teapot asterism's spout. These star clouds are not gravitationally bound clusters but simply regions within our galaxy's disk that are particularity dense with stars, gas, and dust. In contrast, the small Black Swan Cluster, M18, is, by definition, a star cluster: a gravitationally bound grouping of related stars.

OBSERVING LOG

Date:	Time:	Location:

Sky Conditions (Seeing/Transparency):

Instrument:

Eyepiece (Filter/Barlow):

Notes:

Common Name(s): Small Sagittarius Star Cloud (M24),
Black Swan Cluster (M18)
Type: Cloud of Stars (M24),
Open Cluster (M18)
Brightness (Visual Magnitude): 4.6 (M24), 6.9 (M18)
Distance (Light-Years): 10,000 (M24), 4,900 (M18)
Difficulty (Subjective): 2 (M23), 3 (M18)

Observing Tips: From dark skies, the Small Sagittarius Star Cloud can be seen without a telescope. Of course, binoculars and telescopes (at low power) will vastly improve the view. At low magnification, M18 fits into the same field of view as M17 (the Swan). Author and astronomer Stephen O'Meara nicknamed M18 the "Black Swan" after tracing a swan shape with the stars. What do you see?

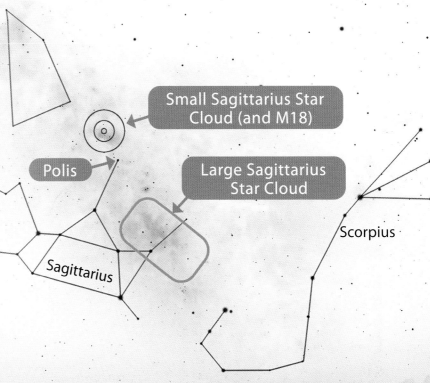

Small Sagittarius Star Cloud (and M18)

Polis

Large Sagittarius Star Cloud

Scorpius

Sagittarius

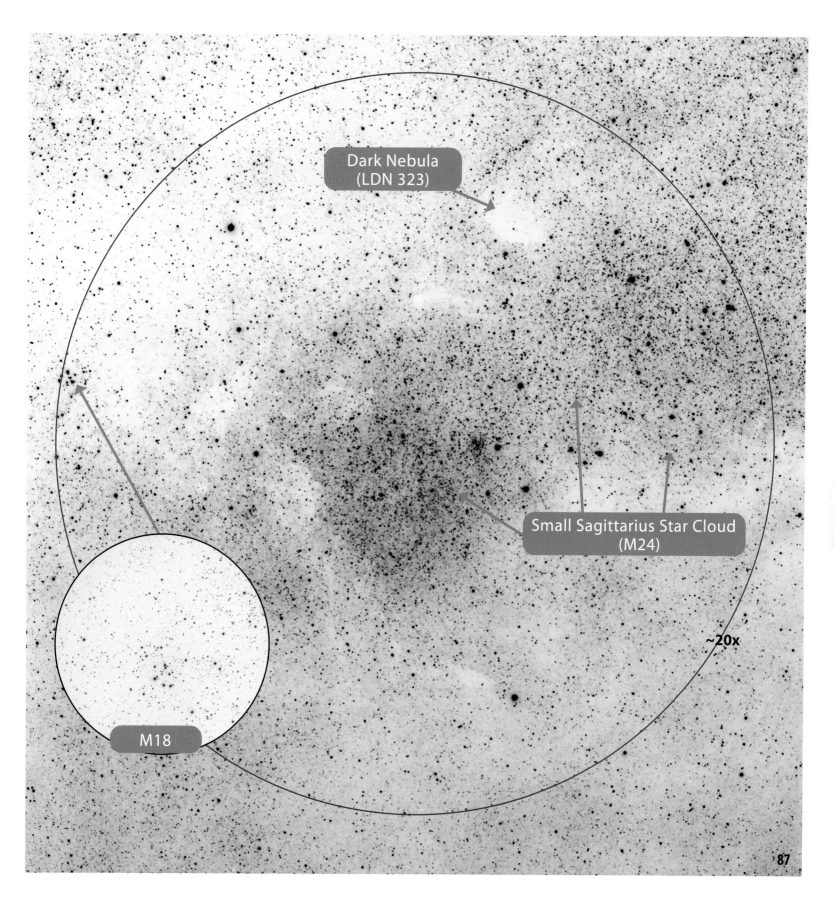

Dark Nebula
(LDN 323)

Small Sagittarius Star Cloud
(M24)

M18

~20x

M23

M23 is a bright open star cluster about the diameter of the full Moon. It is best viewed with binoculars in dark skies. This cluster stands out against a dark background, most likely interstellar dust blocking the light from more distant stars.

John H. Mallas, author of *The Messier Album*, describes this cluster as a bat. The bright star residing near the cluster represents the tip of the bat's tail. John Read, on the other hand, thinks it resembles a stingray.

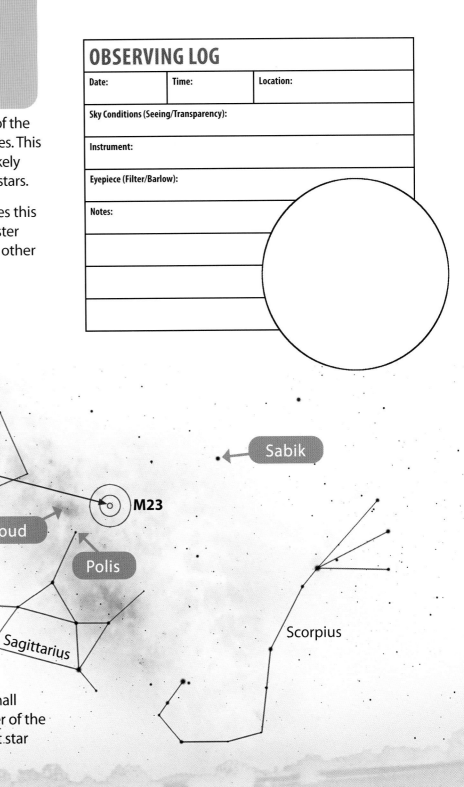

OBSERVING LOG

Date:	Time:	Location:

Sky Conditions (Seeing/Transparency):

Instrument:

Eyepiece (Filter/Barlow):

Notes:

~50x

Sabik

M23

Star Cloud

Polis

Sagittarius

Scorpius

Common Name(s): Mattas's Bat
Type: Open Star Cluster
Brightness (Visual Magnitude): 5.5
Distance (Light-Years): 2,100
Difficulty (Subjective): 2

Observing Tips: M23 sits just to the right of the Small Sagittarius Star Cloud. M23 is also about one quarter of the way along a line drawn from Polis toward the bright star Sabik (a.k.a. Eta Oph).

M25

M25 is an open cluster containing several dozen bright stars, of which about 30 are visible in a small telescope and about 60 in a larger one. Studies show that the cluster contains about 600 stars, most of which are simply too dim for amateur telescopes.

Under extremely dark skies, a moderately sized telescope and averted vision will show "dark nebulosity" running through and around this cluster—clouds of gas and dust that block light from background stars, like ink spilled on a canvas.

Common Name(s): Alligator's Foot (O'Meara)
Type: Open Star Cluster
Brightness (Visual Magnitude): 4.6
Distance (Light-Years): 2,000
Difficulty (Subjective): 3

OBSERVING LOG

Date:	Time:	Location:

Sky Conditions (Seeing/Transparency):

Instrument:

Eyepiece (Filter/Barlow):

Notes:

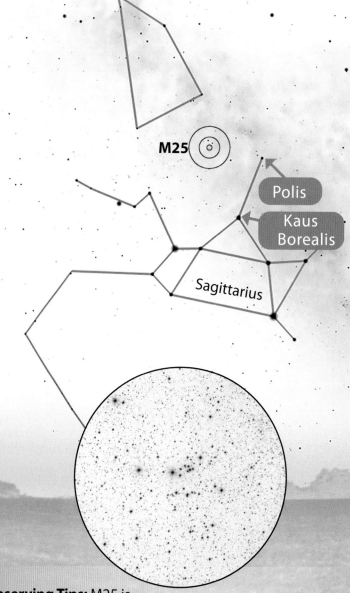

Observing Tips: M25 is found to the left of the Small Sagittarius Star Cloud. From the suburbs, find M25 by forming a lopsided triangle between Polis and Kaus Borealis (a.k.a. Lambda Sgr).

M16

M16 is the target of the most majestic and famous Hubble Space Telescope photograph, the "Pillars of Creation." The eyepiece view photo below was taken by John using the telescope at Saint Mary's University. We didn't invert the image in gray scale in order to preserve the dark towers of gas that make up the pillars.

Don't expect to see the pillars visually. But that's okay—the Eagle's "wings" of glowing gas (which shine brighter in a narrowband filter) and the many embedded young stars will look spectacular under dark skies.

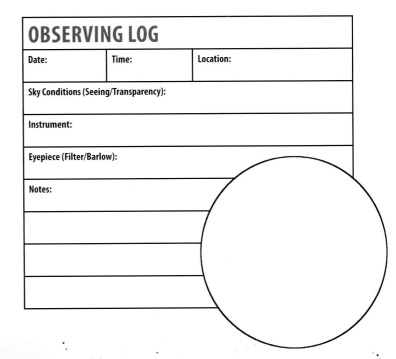

OBSERVING LOG

Date:	Time:	Location:

Sky Conditions (Seeing/Transparency):

Instrument:

Eyepiece (Filter/Barlow):

Notes:

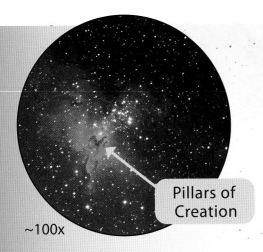

Pillars of Creation

~100x

Common Name(s): Eagle Nebula
Type: Star Forming Nebula
Brightness (Visual Magnitude): 6.0
Distance (Light-Years): 7,000
Difficulty (Subjective): 2

Observing Tips: Although M16 is contained within Serpens, it's easier to extend the bottom edge of Scutum to the upper right, or scan upwards from the Teapot with binoculars on a dark night.

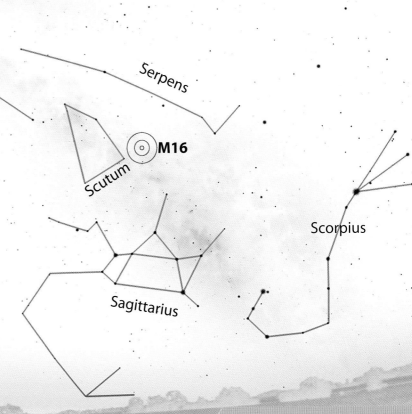

Aquila

Serpens

Scutum

M16

Scorpius

Sagittarius

M17

M17 is found just a couple degrees below M16. It is one of the brightest nebulae visible from the Northern Hemisphere and can be seen in a backyard telescope from the suburbs. It's one of Chris's favorite objects.

The Swan Nebula was given its popular nickname by astronomer George F. Chambers in 1889. When viewing this nebula inverted through a Newtonian telescope, the brighter patches of glowing gas form a serenely floating swan. As your eyes adapt to the dark, and more of the nebula appears, the effect slowly dissipates. A narrowband filter will help bring out the finer details.

Common Name(s): Omega, Swan Nebula
Type: Emission Nebula
Brightness (Visual Magnitude): 6.0
Distance (Light-Years): 5,500
Difficulty (Subjective): 2

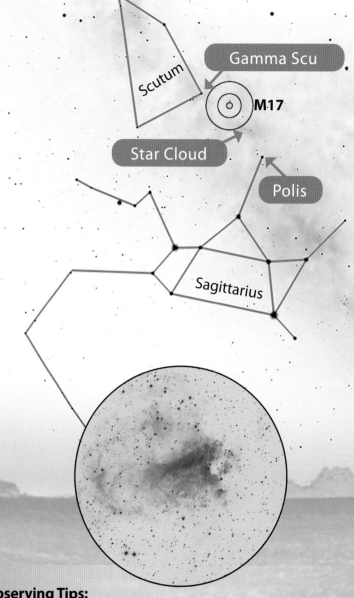

OBSERVING LOG

Date:	Time:	Location:
Sky Conditions (Seeing/Transparency):		
Instrument:		
Eyepiece (Filter/Barlow):		
Notes:		

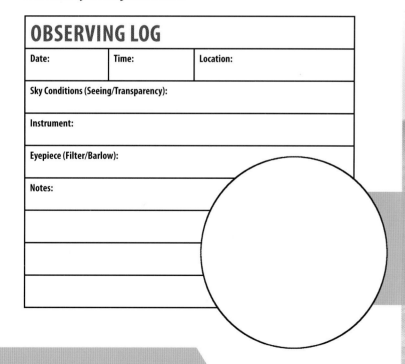

Observing Tips:
M17 sits just above the bright Small Sagittarius Star Cloud (M24), above the imaginary line connecting Gamma Scu and Polis, the northernmost star in Sagittarius.

M11

Although M11 appears almost like a globular cluster (even more so than globular cluster M71 [page 105]), it is an open star cluster, located within the plane of our galaxy. The core of this cluster looks cube-like. For this reason, *Star Trek* fans call this cluster the "Borg Cube" after the cyborg baddies from the popular TV series (there is clearly an overlap between Trekkies and astronomers).

M11 was discovered well before Messier's time and was given its traditional nickname, the Wild Duck Cluster, by Admiral William Henry Smyth, an astronomer (and British Naval Officer), in 1835.

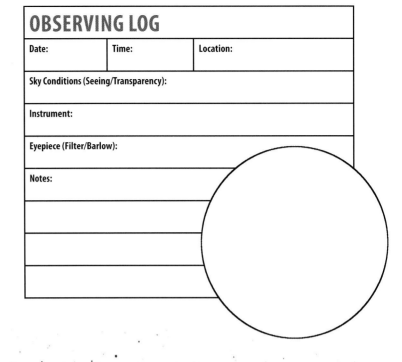

OBSERVING LOG

Date:	Time:	Location:

Sky Conditions (Seeing/Transparency):

Instrument:

Eyepiece (Filter/Barlow):

Notes:

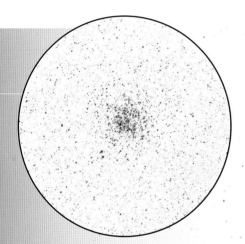

Common Name(s): Wild Duck Cluster, Borg Cube
Type: Open Star Cluster
Brightness (Visual Magnitude): 5.8
Distance (Light-Years): 6,200
Difficulty (Subjective): 2

Observing Tips: The constellation Scutum is fairly dim. When John searches for M11, he imagines the tail end of Aquila as a hockey stick. A few stars at the end of Aquila form the stick's blade, with M11 at the tip of the blade.

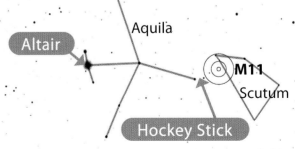

M26

M26 is a small—but dense—open cluster located within the plane of the Milky Way. Its position inside the dim stars of Scutum, and the surrounding rich star-fields, makes it especially difficult to locate. Surprisingly, Messier missed the nearly-as-bright globular cluster NGC 6712 located nearby. From the city, you may be able to locate this cluster through your telescope by simply panning around this part of the sky. The background stars will completely fall victim to the light pollution, whereas the cluster will shine through. Look for a small diamond of four bright stars within the cluster.

Common Name(s): None
Type: Open Star Cluster
Brightness (Visual Magnitude): 8.0
Distance (Light-Years): 5,000
Difficulty (Subjective): 3

OBSERVING LOG

Date:	Time:	Location:
Sky Conditions (Seeing/Transparency):		
Instrument:		
Eyepiece (Filter/Barlow):		
Notes:		

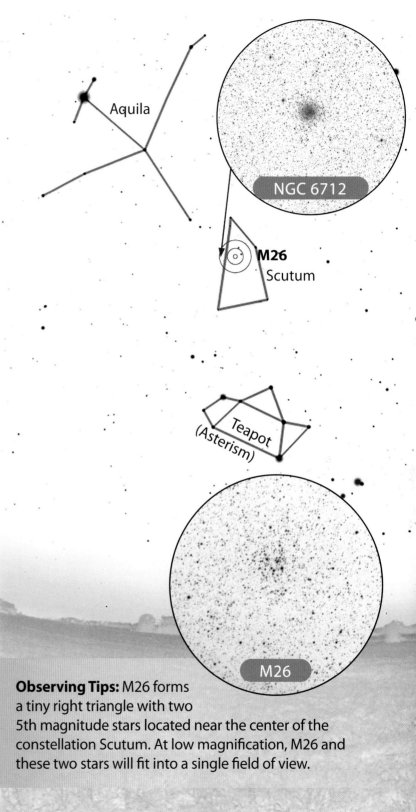

NGC 6712

Aquila

M26
Scutum

Teapot (Asterism)

M26

Observing Tips: M26 forms a tiny right triangle with two 5th magnitude stars located near the center of the constellation Scutum. At low magnification, M26 and these two stars will fit into a single field of view.

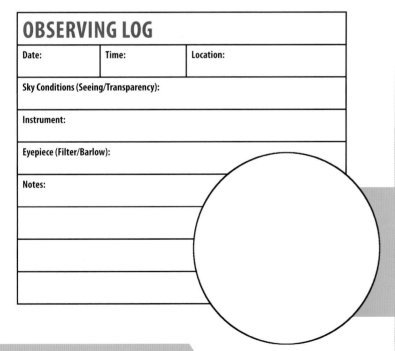

M92

M92 is the little sister of the nearby Great Hercules Star Cluster (facing page)—it's about half the size. The biggest challenge with M92 (and M13) is that they are almost directly overhead in midsummer! Objects that are straight up can be a challenge to observe and track in many telescopes.

M92 is small enough to fit into the David Dunlap Observatory's 74-inch telescope's narrow field of view during Chris's summer public viewing sessions.

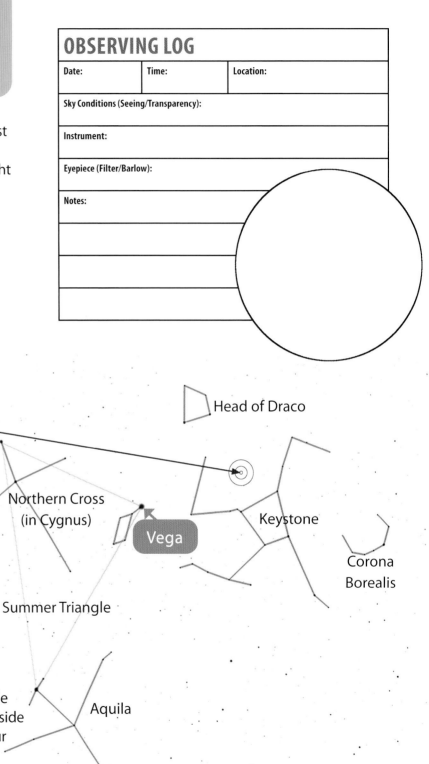

Common Name(s): None
Type: Globular Cluster
Brightness (Visual Magnitude): 6.4
Distance (Light-Years): 26,700
Difficulty (Subjective): 3

Observing Tips: M92 can be found by forming a triangle with the "top" side of the Keystone. Alternatively, it's the side opposite from Corona Borealis, or the side facing the four stars making up Draco's head.

94

M13

M13 is the crowd-pleaser of globular clusters for Northern Hemisphere observers. During summer and autumn star parties, there are always a few telescopes pointed at M13. Binoculars will show it, too! Under dark skies, M13 resembles sugar spilled on black velvet.

When John was a beginner stargazer, this was the first deep-sky object he ever observed. He was using a used 60 mm Meade ETX backpack telescope. Using that little scope, he observed this wonderful cluster from the suburbs just outside San Francisco—once on purpose, and once completely by chance!

Common Name(s): Hercules Star Cluster
Type: Globular Cluster
Brightness (Visual Magnitude): 5.8
Distance (Light-Years): 25,100
Difficulty (Subjective): 2

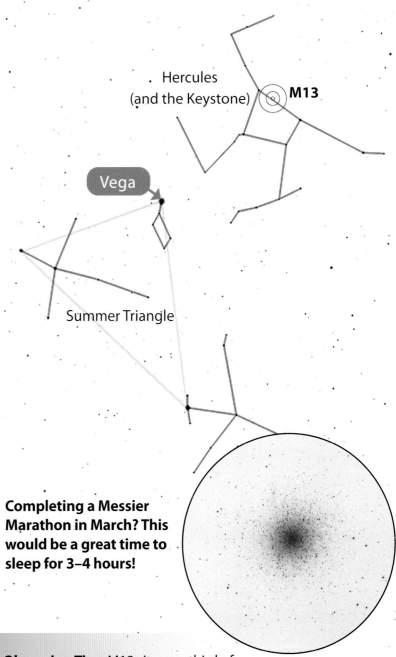

OBSERVING LOG

Date:	Time:	Location:
Sky Conditions (Seeing/Transparency):		
Instrument:		
Eyepiece (Filter/Barlow):		
Notes:		

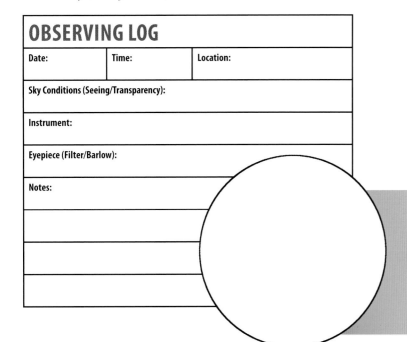

Completing a Messier Marathon in March? This would be a great time to sleep for 3–4 hours!

Observing Tips: M13 sits one-third of the way from the wider end of Hercules's keystone-shaped body, on the side farthest from the bright star Vega. John's method for finding M13 is relatively foolproof, even from the suburbs. He simply pans his telescope around the sides of the Keystone until he runs into it!

M69

Along with M54 (page 100) and M70 (facing page), M69 is the most westerly of three globular clusters found along the bottom of the Teapot asterism—a part of the sky nicknamed "Globular Cluster Alley" by author Stephen O'Meara.

All three of these Messier clusters appear relatively small in the eyepiece at moderate magnification. Even with a larger telescope, you'll need to use high magnification to resolve their individual stars.

Watch for another, smaller, globular cluster, NGC 6652, sitting a finger's width to the lower left of M69.

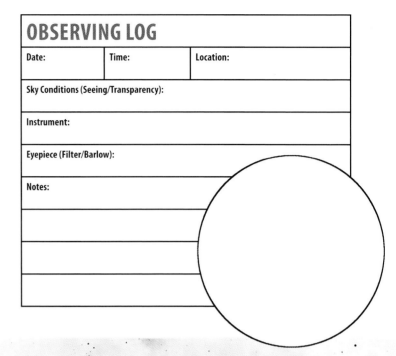

OBSERVING LOG

Date:	Time:	Location:

Sky Conditions (Seeing/Transparency):

Instrument:

Eyepiece (Filter/Barlow):

Notes:

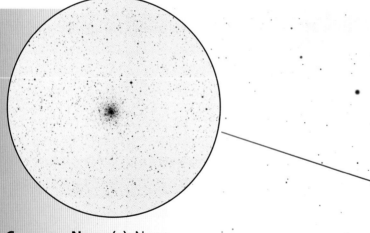

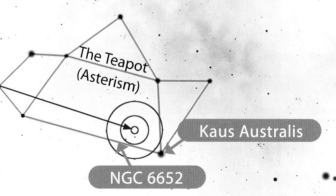

The Teapot (Asterism)

Kaus Australis

NGC 6652

Common Name(s): None
Type: Globular Cluster
Brightness (Visual Magnitude): 7.6
Distance (Light-Years): 29,700
Difficulty (Subjective): 4

Observing Tips: In a dark sky, you should be able to see two stars just to the left of Kaus Australis. From the suburbs they'll be visible through your finderscope. Those two stars and M68 will share your telescope's field of view at about 25x.

M70

M70 is the middle Messier globular in Globular Cluster Alley, and the dimmest of the three.

In Charles Messier's notes, he records that next to M70, "There is a ninth-magnitude star and four faint telescopic stars almost in a straight line." What causes these cascades of stars? Perhaps a cloud of interstellar gas collapsed into a disc, and the stars that later formed within it have not yet drifted out of alignment.

Common Name(s): None
Type: Globular Cluster
Brightness (Visual Magnitude): 8.0
Distance (Light-Years): 29,000
Difficulty (Subjective): 3

OBSERVING LOG

Date:	Time:	Location:
Sky Conditions (Seeing/Transparency):		
Instrument:		
Eyepiece (Filter/Barlow):		
Notes:		

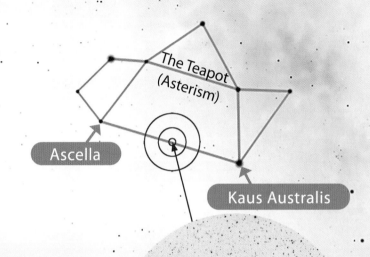

Ascella

The Teapot (Asterism)

Kaus Australis

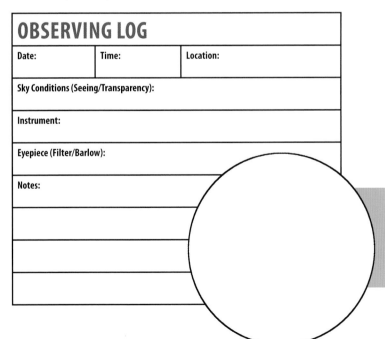

Messier's Quadruplet

Observing Tips: M70 is found exactly midway between the two stars forming the base of the Teapot, Ascella (a.k.a. Zeta Sgr) and Kaus Australis (a wide double star).

M22

M22 is one of the first targets in astronomer Tony Schellinck's "Ace Amateur Astronomer" binoculars program. Though slightly less famous than M13 (page 95), M22 is arguably the most impressive globular cluster visible from mid-northern latitudes.

M22 is slightly brighter than M13, takes up as much sky as the full Moon, and is relatively easy to find. The only challenge for those living up North is that M22 is rather low in the sky, so it's visible for only a few months of the year during evening.

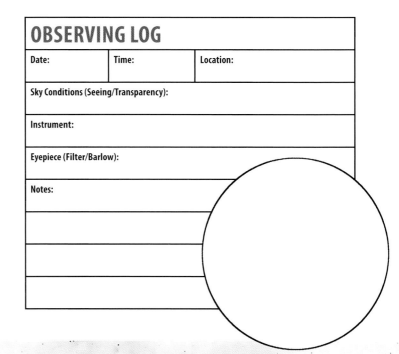

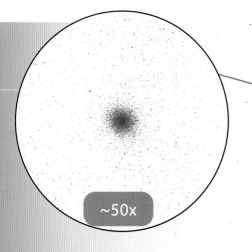

~50x

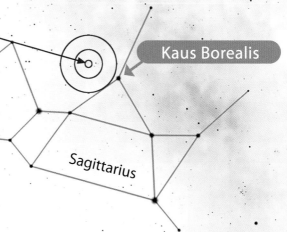

Kaus Borealis

Sagittarius

Common Name(s): Great Sagittarius Star Cluster
Type: Globular Cluster
Brightness (Visual Magnitude): 5.1
Distance (Light-Years): 9,800
Difficulty (Subjective): 2

Observing Tips: M22 is an easy target for binoculars, even from the suburbs. This was the first globular cluster John was able to distinguish without a telescope. M22 is located just left of the top star in the teapot, Kaus Borealis.

M28

M28 is twice as distant as its neighbor M22, and much dimmer. The first millisecond pulsar within a globular cluster was discovered here in 1986 by astronomers at the Lovell Telescope in England. A millisecond pulsar is a small star (specifically a neutron star) that rotates in less than 10 milliseconds, emitting radio waves in our direction with each rotation. On the opposite side of the star Kaus Borealis from M28 sits another small globular cluster, NGC 6638. NGC 6638 is much fainter and was most likely beyond the range of Charles Messier's telescopes, so he missed it in his surveys of the sky.

Common Name(s): None
Type: Globular Cluster
Brightness (Visual Magnitude): 6.8
Distance (Light-Years): 17,900
Difficulty (Subjective): 3

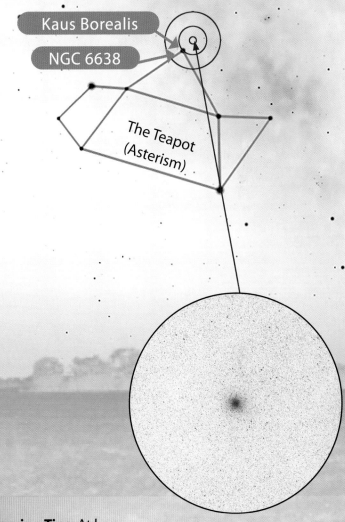

OBSERVING LOG

Date:	Time:	Location:
Sky Conditions (Seeing/Transparency):		
Instrument:		
Eyepiece (Filter/Barlow):		
Notes:		

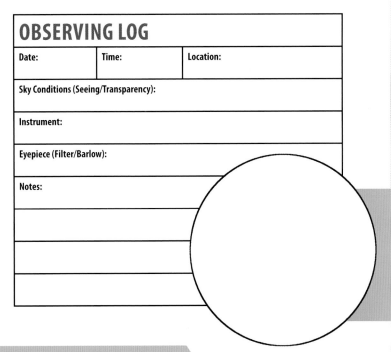

Observing Tips: At low magnification, M28 shares the field of view with Kaus Borealis (a.k.a. Lambda Sagittarii), the star at the top of the Teapot asterism. The cluster is relatively small, and a large aperture telescope is required to resolve individual stars.

M54

M54 is the easternmost cluster in "Globular Cluster Alley." It's also the most distant of all the Messier globular clusters.

Even in larger telescopes, M54 won't look like much more than a smudge. But the cluster is huge! If it were as close as the nearest globular clusters, it would be a spectacular sight in our night sky.

Chris found that this cluster definitely looked larger when he used the averted vision technique.

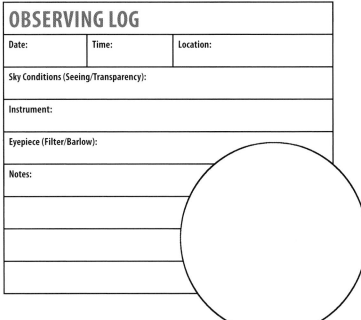

OBSERVING LOG

Date:	Time:	Location:

Sky Conditions (Seeing/Transparency):

Instrument:

Eyepiece (Filter/Barlow):

Notes:

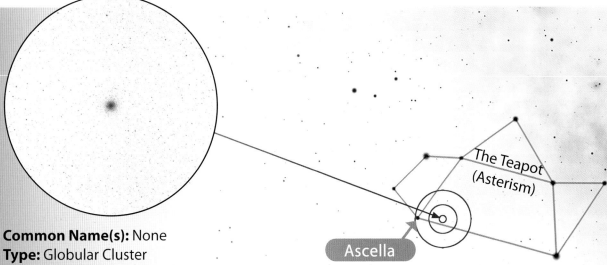

Common Name(s): None
Type: Globular Cluster
Brightness (Visual Magnitude): 7.7
Distance (Light-Years): 87,400
Difficulty (Subjective): 3

Observing Tips: At magnitude 7.7, M54 is the brightest of the three globular clusters in the "Alley"; however, its distance makes it appear small and condensed, so higher magnification is a must! To find it, scan the sky just west of Ascella.

M55

M55 is one of the largest and closest globular clusters. If it climbed higher in the sky at mid-northern latitudes, it would be easier to find and probably be as popular as M13 and M22 (the bright globular clusters in Hercules and Sagittarius, respectively).

From John's site in Nova Scotia, M55 never rises more than 15 degrees above the horizon, which means from most of his observing locations, it's forever hidden behind the trees! From his astronomy club's observatory north of Toronto, Chris was able to see M55 when it was only 12 degrees above the horizon!

Common Name(s): Specter Cluster
Type: Globular Cluster
Brightness (Visual Magnitude): 6.4
Distance (Light-Years): 17,600
Difficulty (Subjective): 3

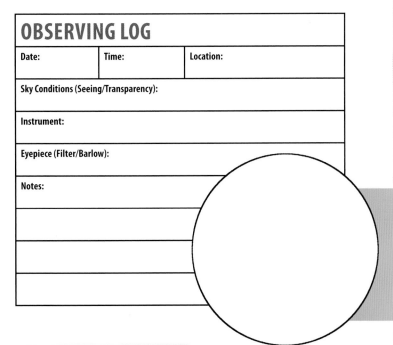

OBSERVING LOG

Date:	Time:	Location:
Sky Conditions (Seeing/Transparency):		
Instrument:		
Eyepiece (Filter/Barlow):		
Notes:		

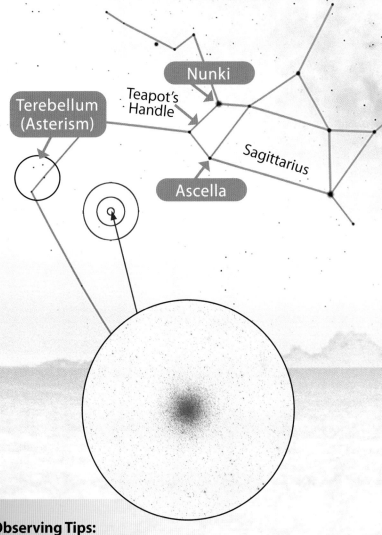

Observing Tips:
To find M55, search the sky to the lower right of the Terebellum asterism (a small cross of bright stars), or extend the Teapot's handle downward from Nunki by almost four times the handle's length.

M56

After observing M13 and M92, this smaller globular cluster might seem like the runt of the litter! But don't let size deceive you; this cluster may be one of the most interesting.

Work up to higher magnification to observe patterns and colors in the stars. This globular cluster is obscured ever so slightly by dust in the plane of our Milky Way Galaxy, giving the stars a unique hue and an irregular shape.

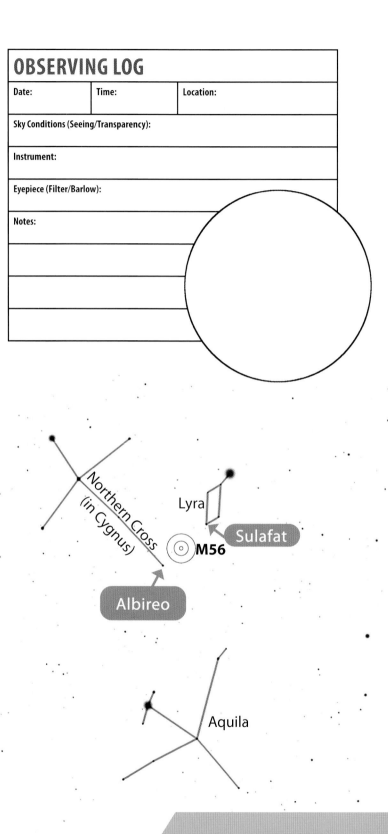

OBSERVING LOG

Date:	Time:	Location:

Sky Conditions (Seeing/Transparency):

Instrument:

Eyepiece (Filter/Barlow):

Notes:

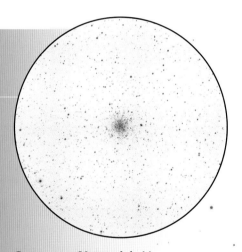

Common Name(s): None
Type: Globular Cluster
Brightness (Visual Magnitude): 8.3
Distance (Light-Years): 32,900
Difficulty (Subjective): 3

Observing Tips: M56 is located almost halfway along a line drawn from the famous double star Albireo to Sulafat, the bottom star in the diamond of Lyra.

M57

After the Pleiades and Orion Nebula, M57, the Ring Nebula, is probably the next most popular Messier object. It may be small, but it's bright enough to be visible even under light polluted skies. We've never had any trouble seeing M57 in any telescope, even from the city.

Even though this planetary nebula will appear as a ghostly smoke ring in your telescope, it's actually a uniform sphere of gas, glowing in the radiation from a tiny white dwarf star shining in its center. Use of a narrowband filter will brighten the ring.

Common Name(s): The Ring Nebula
Type: Planetary Nebula
Brightness (Visual Magnitude): 8.8
Distance (Light-Years): 2,300
Difficulty (Subjective): 2

OBSERVING LOG

Date:	Time:	Location:

Sky Conditions (Seeing/Transparency):

Instrument:

Eyepiece (Filter/Barlow):

Notes:

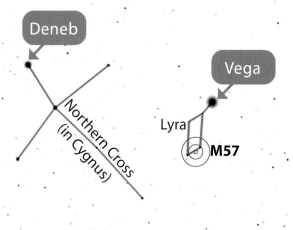

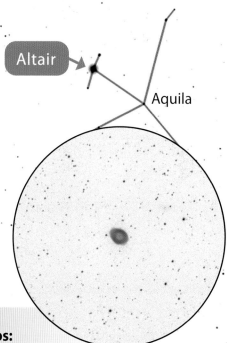

Observing Tips:
M57 is very easy to find.
Aim your telescope midway between the lower two stars in Lyra's diamond. Crank up the magnification to increase contrast and, in a large telescope, try to identify the central star.

M27

M27, more commonly known as the Dumbbell Nebula, looks more like an apple core than its namesake. This is one of the most popular Messier objects to show in telescopes at outreach events because most people have no trouble seeing it clearly.

The Dumbbell Nebula is two to three times larger than M57, the Ring Nebula. It's only one-third as far away, and is estimated to have formed around 10,000 years ago, versus only 1,000 years ago for M57.

OBSERVING LOG

Date:	Time:	Location:

Sky Conditions (Seeing/Transparency):

Instrument:

Eyepiece (Filter/Barlow):

Notes:

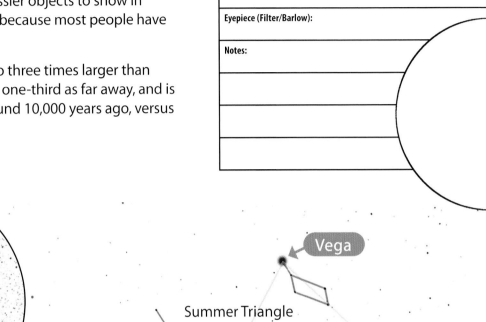

Common Name(s): The Dumbbell Nebula
Type: Planetary Nebula
Brightness (Visual Magnitude): 7.4
Distance (Light-Years): 1,200
Difficulty (Subjective): 2

Observing Tips: M27 is bright enough to observe from the suburbs in most telescopes. A narrowband filter will brighten it nicely. Locate M27 on the northern side of Sagitta's arrow's tip. Alternatively, take a line drawn from the bottom of Lyra's parallelogram through Albireo and double it.

M71

M71 is one of the nearest globular clusters to Earth. This fact, and it's position in a tiny, yet easily identifiable constellation, Sagitta (the Arrow), makes it one of the simplest globular clusters to find.

Though it appears more compact than M13 and contains significantly less stars, M71's closeness makes the individual stars appear brighter.

While you're in this part of the sky, don't forget to check out the "Coathanger" asterism, one of the most popular binoculars targets in the sky.

Common Name(s): Angelfish Cluster
Type: Globular Cluster
Brightness (Visual Magnitude): 8.0
Distance (Light-Years): 13,000
Difficulty (Subjective): 2

OBSERVING LOG

Date:	Time:	Location:
Sky Conditions (Seeing/Transparency):		
Instrument:		
Eyepiece (Filter/Barlow):		
Notes:		

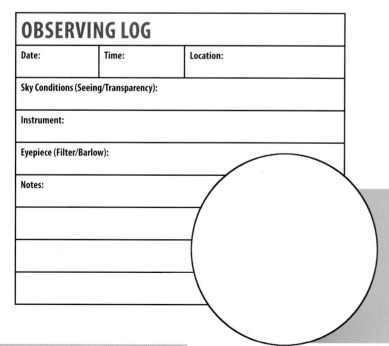

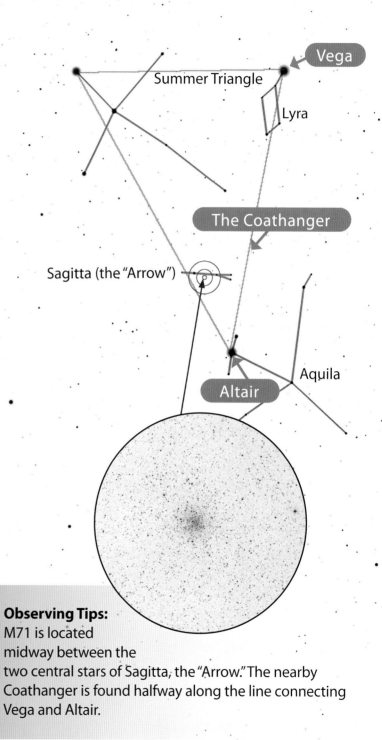

Observing Tips:
M71 is located midway between the two central stars of Sagitta, the "Arrow." The nearby Coathanger is found halfway along the line connecting Vega and Altair.

M29

The open cluster M29 is a favorite of ours. It gets its nickname "the Cooling Tower" from six bright stars, arranged in two rows that arc away from one another. This cluster always reminds John of the huge cooling towers at Springfield's power plant in *The Simpsons*.

There are five Messier objects contained within the Summer Triangle, all of which are visible from the suburbs with a telescope. These are: M57 (the Ring Nebula), M56, M71 (the Angelfish Cluster), M27 (the Dumbbell Nebula), and M29 (the Cooling Tower).

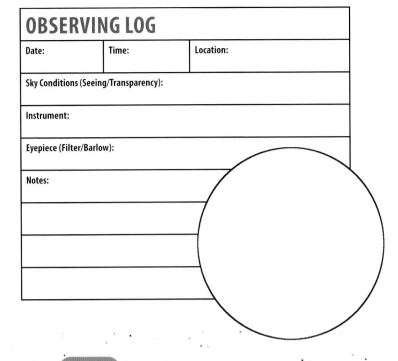

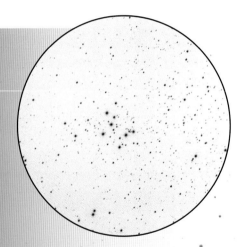

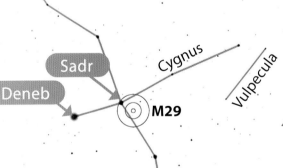

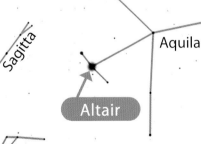

Common Name(s): The Cooling Tower
Type: Open Cluster
Brightness (Visual Magnitude): 6.6
Distance (Light-Years): 5,000
Difficulty (Subjective): 2

Observing Tips: M29 can easily be found in binoculars or a small telescope next to the star Sadr in Cygnus. At low magnification (around 25x and lower), Sadr and M29 will share the same field of view.

This map shows the constellations near the Summer Triangle, an asterism formed by connecting the bright stars Vega, Deneb, and Altair.

M39

M39 is a large, sparse, open cluster near the plane of our Milky Way galaxy. We don't know why Messier never bothered to list more of the many open clusters found in this part of the sky.

In dark skies, this cluster is visible without a telescope, since it covers an area of sky about the size of the full Moon. Binoculars and small telescopes easily resolve most of its member stars, even in light polluted skies. The most challenging part about correctly identifying this cluster is the lack of nearby reference stars, and confusion with other similar clusters in the area.

Common Name(s): None
Type: Open Cluster
Brightness (Visual Magnitude): 4.6
Distance (Light-Years): 825
Difficulty (Subjective): 2

OBSERVING LOG

Date:	Time:	Location:
Sky Conditions (Seeing/Transparency):		
Instrument:		
Eyepiece (Filter/Barlow):		
Notes:		

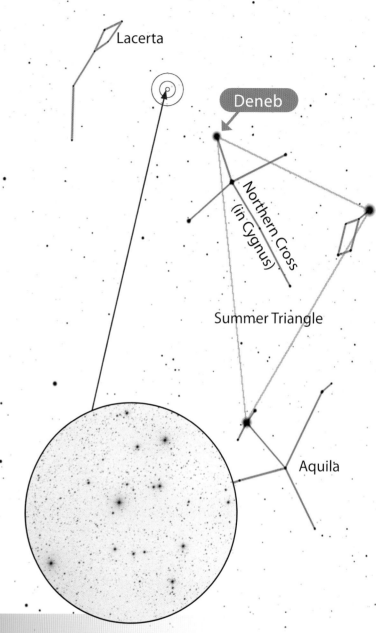

Observing Tips: M39 is located midway between Deneb, the top star in the Northern Cross, and the diamond in the tiny constellation Lacerta. Under light-polluted skies, Lacerta's stars can be hard to see with unaided eyes.

M75

If you've been completing these targets in the order we've listed them in this book, congratulations on reaching your 110th Messier object! If you're completing a Messier marathon, now is the time to return to the remaining autumn targets.

M75 may be the last item in this book, but it only represents one milestone in your stargazing journey. There is so much more to explore, and we hope you attempt even more challenging lists (like RASC's Finest NGCs) and share your stargazing adventure with your community.

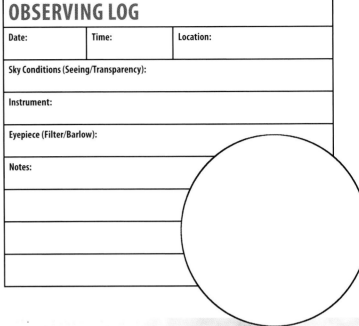

OBSERVING LOG

Date:	Time:	Location:

Sky Conditions (Seeing/Transparency):

Instrument:

Eyepiece (Filter/Barlow):

Notes:

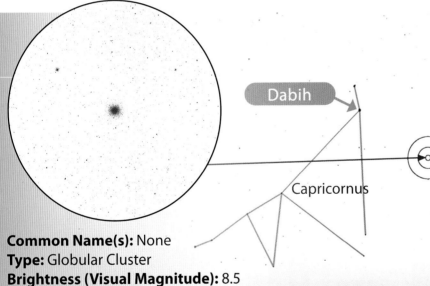

Dabih

Capricornus

Sagittarius

Terebellum (Asterism)

Common Name(s): None
Type: Globular Cluster
Brightness (Visual Magnitude): 8.5
Distance (Light-Years): 67,500
Difficulty (Subjective): 5

Observing Tips: This last target in our list is one of the most challenging. M75 is located a bit more than halfway along a line drawn from Capricorn's brightest star Dabih to the top of the Terabellum asterism. You may want to wait until it's at its highest point (around midnight in the summer) before observing.

Doing a Messier Marathon in March? It's time to complete the remaining autumn targets. M15 (page 6), M72 (page 4), M73 (page 5), M2 (page 2), and M30 (page 3).

Further Reading

***The Messier Objects: Deep Sky Companions* by Stephen James O'Meara.** This is one of the most enlightening books about the Messier List. O'Meara did much of his observing for this book from a dark sky in Hawaii, better conditions than most of us will ever see. He provides the most vivid description of the Messier targets, pushing the limits of human vision.

***The Observing Guide to the Messier Marathon: A Handbook and Atlas* by Don Machholz**. Machholz's book presents several regimented approaches to completing the Messier list in record time. This book contains extensive tables and charts designed for the most diligent observers.

***The Year-Round Messier Marathon Field Guide* by H. C. Pennington.** Pennington presents a very similar approach to finding the Messier objects as this book. His book includes finderscope views (whereas this book provides Telrad rings instead), constellation maps, and sketches for each of the items in the Messier Catalog.

*As of this writing, some of these books may be out of print. If you cannot find these resources online, be sure to check your local library or local astronomy club.

Hubble's Messier Catalog: See the Messier objects as imaged by the Hubble Space Telescope. https://www.nasa.gov/content/goddard/hubble-s-messier-catalog

RASC Visual Observing Programs: Completed the Messier program? Find your next certificate program here. https://www.rasc.ca/certificate-programs

Astrospheric—Astronomy Weather Forecasting: This app shows clouds, sunset, moonset, seeing, and transparency in one simple forecasting tool.

Stellarium: The best free stargazing software. This app is available for PC, Mac, iOS, and Android. A web-based interface is also available.

SkySafari 7 Pro: One of the best (paid) astronomy applications. Not only can it be used to find objects in the sky, but it can also be used to control your computerized telescope!

RASC Certification

RASC Observing Program Application for Certificate

Congratulations on completing the RASC Observing Program!

Program: Messier Catalogue - Traditional ☐ Go-To ☐ (Check one)

In order to receive your certificate, please complete this form and submit it to:

Chair, Observing Committee
Royal Astronomical of Society of Canada
203-4920 Dundas St W
Toronto, ON, M9A 1B7
Email: *observing@rasc.ca*

Affidavit:	**Authentication:**
I, _____ attest to the following:	We declare that _____
I have observed all the features of the *RASC Observing* program and I have (1) included a copy of my logbook (or other record of my detailed observations), including: date, time, location, instrument, magnification, brief description, and a sketch (optional) or photograph (optional) **OR** (2) had my detailed observations authenticated by two RASC witnesses, identified to the right.	has qualified for the *RASC Observing Program* Certificate based on our review of the detailed observation records provided, as well as our personal knowledge of the applicant.
	Witness 1 _____ Date_____
	RASC Centre _____
	Email _____
Name _____ Date: _____	Signature _____
Address _____	
Address _____	Witness 1 _____ Date_____
E-Mail _____	RASC Centre_____
Certificate Language	Email _____
Centre of Affiliation	Signature _____
NOTE: The applicant must be a member in good standing of the RASC to be awarded this certificate.	

110

Astronomical League Certification

- This certification is available to members of the Astronomical League.
- Observers completing 70 Messier objects will receive a certificate.
- Observers completing 110 objects will receive an honorary certificate and pin.

Your notes must include:

1. Date and time of observations (local time or UT)
2. Location of observations (or Latitude and Longitude)
3. Seeing (rated 1 [best] through 5 [worst])
4. Transparency (measured by the magnitude of the faintest star you can see without a telescope.
5. Aperture of your telescope
6. Power [magnification]
7. A description of the Messier object as it appears in the eyepiece.

Additional Rules:

- No automated or go-to telescopes, or any other type of assistance such as apps that have a pointing function. Only star hopping with maps (such as those in this book) and your finder are acceptable for locating targets.
- No reusing observations from other programs (such as binocular programs).
- No observations from marathons or sprint sessions. Sufficient time must be taken to appreciate each Messier object.

To receive your certificate (and pin):

- Your observation log must be examined by an officer of your Astronomical Society or a suitably qualified second party (such as a seasoned member of your local astronomy club).

Complete rules found here:

https://www.astroleague.org/al/obsclubs/messier/mess.html

Make your notes unique

The intent of the object description is to pick out each object's unique details to the best of your ability.

- Is the object round, oval, square, or irregular shaped?
- Does the object have sharp edges, or does it fade away?
- Does averted vision seem to increase its size or help reveal more details?
- Does the object have a bright core?
- In open clusters, can you estimate the number of visible stars?
- In nebulae, are some parts brighter than others?
- Does the nebulae contain any interesting patches of stars?
- What else is in the field of view that is interesting? Is it densely packed with stars? Did a satellite just pass through? Etc.

INDEX IN MESSIER ORDER

SOURCES

Technical data source: The SEDS Messier Catalog Webpages by Hartmut Frommert, Christine Kronberg, Guy McArthur, and Mark Elowitz. SEDS, University of Arizona Chapter, Tucson, Arizona, 1994—2018. http://messier.seds.org/ (Although some updated sources have been used.)

The RASC's *Observer's Handbook*, 2021 edition.

General information on Peak Surface Brightness by Tony Flanders: https://tonyflanders.wordpress.com/messier-guide-index-by-number/

Image Credits

100x eyepiece view images were taken by the author using the Burke-Gaffney Observatory and the Sierra Stars Observatory Network.

Certain wide-field images courtesy of the Digital Sky Survey (DSS).

30–50x eyepiece view images taken by the author using various instruments.

Star maps used in this book were sourced using Stellarium, an open-source stargazing program. These maps were then customized for the purpose of this book. Stellarium is the best free astronomy software out there. A link to this software can be found here: http://stellarium.org.

Image of Charles Messier, Public Domain, Stoyan R. et al. *Atlas of the Messier Objects: Highlights of the Deep Sky*. Cambridge: Cambridge University Press, 2008, p. 15.

"Seeing" image credit: Canadian Meteorological Centre

ACKNOWLEDGMENTS

I'm always surprised how many people it takes to produce a book. When asked to help with this project, the following people were overwhelmingly generous with their time and experience: A huge thank you to Chris Vaughan for agreeing to coauthor this book with me. To Blake Nancarrow from the Royal Astronomical Society of Canada's Observing Committee for offering advice on how to design this book to be used in the RASC's observing programs. To Scott Kranz for providing instructions on the Astronomical League's Messier certification requirements. To proofreaders, and executive consultants, Dr. David Hoskin, David M. F. Chapman, and Kara Turner. To *Star Trek: Voyager*'s Tim Russ for the inspirational foreword. To Ray Khan of Khan Telescopes for tracking down a Sharpstar 61 during the 2021 global telescope shortage. And to my wonderful wife Jennifer, for giving this book its beautiful and intuitive interior design, and for keeping our three kids at bay as we finalized this book during the COVID-19 lockdown.

NOTES

NOTES

NOTES

NOTES